Building Mobile Experiences

Building Mobile Experiences

Frank Bentley and Edward Barrett

The MIT Press
Cambridge, Massachusetts
London, England

MIT Press books may be purchased at special quantity discounts for business or sales promotional use. For information, please email special_sales@mitpress.mit.edu.

This book was set in Stone Sans and Stone Serif by Toppan Best-set Premedia Limited, Hong Kong. Printed and bound in the United States of America.

Library of Congress Cataloging-in-Publication Data

Bentley, Frank, 1979–
Building mobile experiences / Frank Bentley and Edward Barrett.
    p. cm.
Includes bibliographical references and index.
ISBN 978-0-262-01793-0 (hardcover: alk. paper)
1. Mobile computing. 2. User interfaces (Computer systems) 3. Mobile communication systems—Equipment and supplies. 4. Application software—Development. I. Barrett, Edward. II. Title.
QA76.59.B427   2012
004—dc23
                                                                          2012004563

10 9 8 7 6 5 4 3 2

Contents

1 Introduction

The mobile device, more than any other recent invention, is dramatically changing the ways in which we interact with each other and with our cities. Mobile apps, and the resulting mobile experience, are radically different from the fixed-location desktop or laptop experience and require a different design philosophy. Just how they differ, and how our approach to the design and creation of these experiences is different, will be the subject of this book. We are not showing a way to create a quick app for the moment. We are discussing an approach to designing mobile media that answers the ineffable human need to sense, communicate, and understand our own and others' experience of being in a real place in real time, an experience that is simultaneously private and social, and mediated by various forms of technology. We are interested in creating experiences that fundamentally change the ways in which people communicate, are aware of others, and engage with their social networks, cities, and governments: experiences that empower people to be better friends, neighbors, and citizens.

When we use the word "experience," we aim to move beyond current practice, where the word sometimes substitutes for user interface. To us, a mobile experience is everything that happens to a person once they learn about a new application. While the details of the user interface on the device are important, the experience includes all of the side actions and real-world interactions that are enabled by that new mobile application or service. Building an experience means building systems that support authentic, real-world interactions: interactions with people and places that change the way users experience a place and time.

The mobile phone is the ideal platform for these rich, contextual experiences. Already, mobile phones have changed the lives of the majority of people in the world. Think back to your life before a mobile telephone. If you were running late, the only way to tell someone was to find a pay phone. And that only worked if the person you needed to contact was in a location that had a fixed phone, such as home or work, and if you knew their phone number in that location. If you were out and needed directions, you had to buy a paper map or ask someone on the street how to get to your destination. And if you wanted to capture a photo to share with someone, it required a camera, some cables, a computer, and a connection to the Internet. It was a world that is hardly recognizable to anyone under twenty living almost anywhere in the world today.

Now think of your interactions with a computer. These interactions occur in a fixed space and usually for much longer durations than your brief interactions with a phone. Computer-based tasks have less to do with your immediate surroundings and more to do with a task that you are trying to accomplish, whether it is creating a report or sending an email to your mother. To use a computer on the Internet, you need to find a location with Wi-Fi access, generally at home, work, or a fixed third place like a coffee shop. This has made studying computer-based interactions fairly easy, while the mobility and contextual dependence in mobile phone interactions bring new challenges to designing, building, and evaluating new mobile experiences.

The mobile phone has dramatically changed the way that people communicate with each other and with their environments. Over the past two decades, mobile devices have transformed from simple mobile telephones to Internet-connected, context-aware, media capture devices capable of a new set of functionality not present in a traditional computer. These changes have created new industries and hundreds of thousands of commercial mobile applications. While this change is still in progress for many, it is a change many researchers saw coming years ago. Over this time, we have refined a set of methods that have proved useful in building systems that take advantage of these differences in use.

Communities of researchers in Ubiquitous and Pervasive Computing have been creating and studying these types of devices and interactions since the 1980s. But not until the last several years have we had the mass-market devices to actually build and deploy many of the applications that have been dreamed of in limited research contexts. And in this process of designing and building these experiences, we have learned many lessons on the appropriate methods to use in each of the steps of creating a new mobile application or service. We have learned that creating a mobile application that will greatly impact people's lives and their communication requires some changes in process from traditional web or PC development.

This toolkit of methods is the main topic of this book. The process we will explore was developed over many years of mobile research at Motorola and MIT. It takes the best of existing research and product design methods and adapts them for a mobile setting, where the way in which a new application is used takes advantage of the unique properties of the mobile device and information about its environment. We will explore the new mobile paradigm, from both novel functionality and interaction perspectives, by following several successful projects from Motorola Labs and a class we jointly teach at MIT. Throughout the book, we will address the methods we applied to invent compelling concepts and how we took them through research, design, and in many cases to successful commercial products.

At a high level, this book is about the user-centered design process applied to mobile. Many of the methods and the overall process of rapid iterations while building fidelity and getting feedback from real users will look familiar to anyone who has been involved in the design or testing of almost any digital system in the past fifteen years. However, we will highlight the changes to these methods that help make them more relevant to the mobile design process and to the use of systems in the world instead of at a desk. We hope that this book will leave the reader with a large toolbox of methods that can be applied at all parts of the mobile design and development cycle to help create experiences that fit into the complex realities of daily life.

By starting with a deep understanding of people and their behaviors in a domain of interest and by constantly observing the use of early prototypes in real people's daily lives, this design process leads to new experiences that work with how people want to live. We will show how groups at Motorola and MIT have consistently invented compelling, patented, and commercially successful systems when starting from generative research studies with a diverse set of people in the world. None of these concepts would have been possible without the insights gained from this research, and certainly the details of each interaction would have been lacking if not for a deep understanding of current practices and the desires of potential users. In discussing each phase of research, we will clearly describe how to get the most out of that step in the process to better understand the rich ways in which people are experiencing life or would like to experience life, and how to turn these insights into successful product concepts.

Successfully applying this process in a product cycle is not an easy task. Finding documented examples of products that began with generative research in a domain of interest, were built up through rapid iterative prototyping and user evaluations, and finally made their way to market as successful products is difficult. We hope that the examples in this book will help to inspire others to follow the process from start to finish and that they will end up with better products as a result. Along the way we will provide some shortcuts and discount methods for those who do not have the luxury of time often afforded by an industry research lab or university setting.

Before diving into this process, it is important to explore those characteristics that make the mobile device a unique piece of technology. We will explore three main features of a modern mobile device that make it quite different from a standard computer. These are its abilities to function as an always-accessible instant social connection to the people in our lives, to capture its environment, and to sense its surroundings. We will explore each of these topics throughout the remainder of this chapter. The remainder of the book will explore the implications of these new functionalities to the ways that we design and build new experiences for mobility.

Social Connection

When mobile phones were first commercialized in the late 1970s and early 1980s, they were fairly simple devices. Looking like a large brick with a one line numeric display and a number pad much like the Bell home telephones of the time, they were created to make voice telephone calls and nothing else. Even with this basic function, the ability to place and receive a phone call from anywhere (where there was a signal) brought great new freedoms to communication.

Initially, most cell phone users were doctors and professionals who had a need to be accessible while not in front of a landline telephone. Mobile networks were mostly limited to dense urban areas and coverage indoors was spotty at best. For many people at the time, the mobile became yet a third phone in their lives for the spaces in-between home and work.

Figure 1.1
Evolution of mobile phones. Phones have evolved to give us richer ways to communicate with friends and family. From large bag phones, to portable devices, to text and picture messaging, to smartphones with full email and web capabilities, the opportunities to connect to others with a mobile device have been increasing rapidly. (Images used with permission of Motorola Mobility)

Over the years, as the devices became smaller and the networks expanded to cover greater areas, the mobile device became a truly ubiquitous mode of communication. Mobile phones became the default way for most people in the world to get in touch with each other. Simply knowing that most people carried their phones on them led to a change in how people communicated, planned, and met in person. These changes were highlighted by social science researchers Ling and Yttri (1999) in what they termed "Micro-coordination." No longer did people have to make precise plans when meeting up. They could agree on an approximate location and time and refine the plan as time went along by calling or texting each other on their mobile phones. The rise of "third places" like coffee shops and shopping malls helped to give people places to wait when plans changed (Oldenberg 1989). As one of Ling's participants told him, "No one sits at home and waits for the telephone to ring."

As everyone started to carry a mobile phone, the device morphed to become a part of one's body. Phone numbers became associated with people and not places, as had been the case with landline telephones. Before mobile telephones, a home phone number would be shared by a family and often a person would have multiple numbers for the different places in their lives. Mobile telephones changed this and created a number that was truly a connection to a specific person, almost anytime, and almost anywhere.

With this change brought dramatic changes to communication. The ability to instantly be in touch with anyone over a distance worked to change the perceptions of distance itself. With the end of the concept of long-distance charges on mobile phones in America and the fact that people started carrying their devices with them, the ability to call someone over hundreds or thousands of miles and instantly talk to them wherever they were became commonplace. It seems so natural to us today, but just fifteen years ago this was not common practice and long-distance calls were seen by many as a costly event for a special occasion.

As technology progressed, new forms of communication became possible. The Short Message Service (SMS) provided people a means to send a short 160-character message directly to another person. With text messaging, mobile devices became truly personal communications technologies. Mobile phone conversations are by nature public and by social convention usually longer than a few sentences. Text messaging brought along a way to communicate with a person in another location without the need for a full conversation or interrupting others around the caller with details of their private lives. In addition, the received messages could be read and responded to asynchronously, further allowing for tailoring the communication to the availability of the recipient.

Studies of SMS use, especially with teens, have shown the use of this service for a totally new form of communication. SMS allows communication to occur in the background, often while the user is occupied with another task or is in an area where it is not possible to speak out loud. Around the world, SMS began to be used to maintain a constant awareness of close friends and family with short updates being sent to indicate that someone was running late or just that they were thinking of the recipient. Many teens started sending and receiving thousands of these short messages each month as a way to stay connected to their social networks. The work of Grinter, Palen, and Eldridge (2006) has studied the use of SMS in creating greater independence for teenagers and the ways in which families use SMS to check in with each other when apart. This has helped some families to reverse the trend of decreased autonomy for teenagers.

As Internet data services began appearing on mobile phones in the early 2000s, almost all forms of mediated communication became available on this mobile platform. Email, instant messaging, voice over Internet services like Skype, and live video chat came to many devices. In making the transition to mobile devices, these services fundamentally changed. Email, a service once tied to a place, much like the telephone, transformed into a means to access a person no matter where they were. Distinctions between home and work continued to blur more than ever before as greater numbers of people were expected to be always connected to their latest work emails from their devices.

Numerous studies have observed the addiction of mobile-email users to their devices. The term "Crackberry" became a common term to refer to this addiction, being recognized as Webster Dictionary's word of the year in 2006 (Santora). This obsessive checking of email and other web-based data was not limited to the Blackberry devices. Similar use is seen on other smartphones that are now on the market. Finding solutions to prioritize interruptions is an active area of research.

As mobile data services and unlimited data plans became increasingly common, new types of instant communication became possible. One of the case studies in this book will follow the design of the social phonebook for MotoBLUR (Bentley et al. 2010), a push social-networking service that keeps the user up to date on all of their friends' updates to online social networking sites. Mobile presence services like this bring a new awareness to the current activities of friends and family, from tweets to status updates to photos posted to the web. Seeing these in real time serves to remind users of those that they care about most and helps to find opportunistic times to meet in person or talk about updates that inspire communication. Different from other types of interaction, this information is all pushed to the device, so it is always accessible from one place, instead of forcing the user to obsessively check many different websites or applications to find out about their friends.

Today, the mobile device serves as the default connection to people in our lives. It is an always-on connection straight to the people we care about and is evolving to be a platform that keeps us constantly aware of each other through instantly shared text, voice, photos, and videos. This interaction is quite different from how people interact with their computers, which usually occurs in longer bursts throughout the day and in particular locations each time. The anytime, anyplace, quick interactions on a phone typified by the brief nature of a text message requires a different style of design, research, and evaluation—one that can design for use in a wide variety of situations and evaluate the usefulness and usability of an experience in those real-life situations outside of a lab. Mobile systems enable connections to particular people or information in particular contexts with particular constraints. This always-available connection to others is a property of carrying a portable, always-on device. The other two features of modern mobile devices—sensing and capturing environments—rely on the particular hardware that is present in today's new mobile devices.

Live Environmental Capture

Up until the early 2000s, the only way to share a live real-world experience with someone else through a mobile phone was through the audio of a phone call. Concert-goers could hold up a phone for others to hear their environments and share, at least a little bit, in that experience from a distance. Others chose to describe their surroundings to each other through words. In a study Frank and his colleague Crysta Metcalf conducted at Motorola Labs in 2005, we had participants record their mobile phone conversations for us (2008). Among other things, we listened for how our participants used words in a phone call to share details of their current environments with others. Participants discussed modes of transportation they were in, buildings in their sight, and activities going on around them. They talked about seeing a "low flying plane" in the city, or gave contextual awareness such as "I'm on the train now, I wasn't before." In these cases, the phone was serving as a fairly basic tool for these people to share a bit of their experience in the world with others.

Starting in 2003, mobile phones started including built-in cameras. These early cameras were fairly primitive in terms of resolution, but they opened up the possibility for instantly sharing visual scenes with others from any place that had a signal. However, this sharing was plagued with issues. Multimedia Messaging (MMS) initially had problems in sending messages across different mobile operators, and (at least in the United States) both the sender and receiver were charged up to $1.00 per message. This led to a general lack of adoption of MMS services with only 6 percent of mobile users in the United States using the service in 2005 despite popular phones from many manufacturers supporting the feature (Blackwell 2006).

However, as phones began to support Internet data service and initial mobile applications appeared, written in a small subset of Java called J2ME or natively for the Symbian platform, the options for sharing photos greatly expanded. Services like Radar.net and Yahoo! Research Berkeley's ZoneTag, which built on research from UC Berkeley and Stanford, allowed for easy sharing of photos to communities of friends on their respective websites. This enabled a new type of instant sharing of one's immediate surroundings with friends and family through mobile devices. This sharing transformed the device from simply a medium to share words to a rich platform to share a visual experience instantly with others. A photo sharing service named Radar created a new type of "lightweight visual communication," which was researcher Maia Garau's way of describing the ability to share and see photos from others as a natural part of the day instead of only on special occasions and in person (Garau et al. 2006). With these mobile systems, sharing a photo was possible all on one device, with just a few key presses. This was a large improvement over previous photo sharing, which involved digital cameras, cables, laptops, and email.

ZoneTag, as another early example from 2006, posted photos from mobile phones directly to the Flickr website for sharing with already established friends and family (Naaman and Nair 2008). On select Nokia and Motorola devices, users could take a photo, tag it with contextually meaningful labels, and instantly share it to their friends on Flickr (Ames and Naaman 2007). Friends could see the photos on the Flickr website or through a special-purpose mobile photo viewer called Zurfr. Zurfr, as well as the Motorola version of ZoneTag, let users see a current feed of friends' photos as a means to be aware of visually interesting parts of their life (Naaman, Nair, and Kaplun 2008). The applications also showed recent photos taken by anyone nearby as a way to understand one's surroundings. This instant photo sharing opened up the world a person can see to anyone who happened to be in the area.

In the early years of camera phones, several high-profile incidents worldwide illustrated the value of Internet-enabled cameras in a great many people's pockets (Noguchi 2005). During the terrorist attacks in the London Underground (Borenstein 2005), hurricane Katrina, and recent unrest in the Middle East (Grossman 2009), mobile photos and videos have opened the images of these tragedies to the world and have helped the professional media as well as government services better understand and respond to unexpected situations (Liu et al. 2008).

Another area of applications that draws on the community to submit photos and other aspects of their environment is called Participant Sensing Systems. In one example from Deborah Estrin's group at UCLA (Reddy et al. 2008), citizens can capture pictures of broken sidewalks in their communities and upload them to the local government so that they can be added to a list to be repaired. This type of system is much easier and more reliable than needing to call a government office and verbally describe the condition of the hazard to another person who must then relay it to a city inspector. These systems can engage citizens in new ways in the context of their cities and communities and enable government to be more efficient in responding to the needs of its residents.

Throughout the 2000s, mobile cameras continued to improve in quality. By 2006, typical mobile devices brought 1.3 megapixel cameras and video recording capabilities. However, without installing custom applications, it was still difficult to share with others the images and video that were captured. This would soon change with mobile devices that offered rich application platforms and mobile email.

Research in Motion's Blackberry and the Apple iPhone lowered the bar to sharing mobile images with others. Both platforms made it simple for users to attach images to emails and send them along at no additional charge and, more importantly, at no charge to the receiver. These platforms also began to support rich Internet applications to allow easy upload of content to sites like Facebook and Yahoo's Flickr photo sharing platform. Various application stores from manufacturers and operators made these programs easily discoverable and provided a simple download process that dramatically increased application use from the days of J2ME applications, which were usually hosted stand-alone on the web pages of individual sites, and each application had a different process for downloading to a device.

Beyond photographs, systems that allow live video streaming are now gaining in popularity. At the time of this writing, the two most popular are Bambuser (2011) and Qik (2011). These services allow users to stream live video to the web to allow anyone to see a live view of exactly what they are seeing. The power of live video streams, whether to the world though systems like this or in person-to-person contexts such as through mobile Skype video or Apple's FaceTime, provides unparalleled abilities to share our surroundings and experiences with others over a distance. As Dougherty (2010) has studied, these systems are being used for social activism and awareness by users around the world. She found that 11 percent of the videos uploaded to the Qik site had civic value, with many videos having journalistic or educational value.

Today, it is easier than ever before to share the sights and sounds of our worlds with people we care about and with the greater public world. With an estimated 227 billion camera phone images captured in 2009 (InfoTrends 2011), there is a large potential to share what we see with others. In addition to sharing with others, these images are also being used for ourselves. Camera phones allow people to capture photos of products they wish to buy, document events such as car crashes, and capture little bits of their day that have never before been recorded. Because this capture, sharing, and receiving occurs out in the world, a new set of methods is required to design and study the use of mobile media applications. Studying use in a lab will not highlight the real use in practice of these contextually rich applications or the problems that users experience in the real world. Nor will they capture the rich emotions that images from the lives of our close friends and family evoke when shared and received instantly through a mobile device.

The latest set of mobile devices is capturing more than just the sights and sounds of the world. Phones and portable media players are integrating with the body and becoming wearable computing sensors to track our workouts and monitor our health. These types of applications will be increasingly prevalent as it becomes easier to connect sensors to phones and run monitoring applications in the background while doing other tasks. These systems move beyond simple capturing of the visual/auditory environment to capturing the rich sensory context of the user.

Contextual Sensing

As soon as phones became portable, the location of the user became unknown to the caller. Thus, "Where are you?" became one of the most used cell phone greetings. The American carrier Boost Mobile (2011) even used "Where you at?" as a marketing slogan. Context is important to a phone call to help establish how long someone is available for communication, what topics are in-play to discuss given their current surroundings, and in helping to coordinate logistics. As soon as mobile devices began to carry sensors, the ability to share context and use it to help with navigation and information seeking began to appear in many applications.

One of the first contextual sensors to appear on mobile phones was location. As a cell phone travels throughout the world, it constantly hands off its connection from one cell tower in the network to another. These towers can be mapped, and a user's approximate location can be determined based on the tower that their phone is connected to and can be refined if the phone has information about all of the cell towers in view and their relative signal strength (Wortham 2009). Researchers at Intel and Yahoo! created databases of these mappings in the mid-2000s, but they grew at a fairly slow pace as users had to identify cells one at a time. Yahoo! Research Berkeley's Zone

Tag application, mentioned above, aimed to build this database faster as users had the incentive of receiving recommended tags for photos if their current cell was labeled with a City, State, and Zip Code in the ZoneTag database (Ames and Naaman 2007). These recommended tags came from several sources including Yahoo!'s existing database of places called Yahoo! Local. So if a user took a photo in Millennium Park in Chicago, suggested tags might be "Millennium Park," "Cloud Gate," "Pritzker Pavilion," "Grant Park Orchestra," etc. based on places and events that existed in various Yahoo! systems as well as tags that other users had contributed in that location in the past. However, this could only work if Yahoo! had received a mapping of the user's cell tower identification to an actual place.

As Global Positioning System (GPS) sensors came to mobile phones, both Yahoo! and Google extended their location-aware applications to correlate GPS position with cell tower information to help build these databases automatically and at a much faster pace. Any time a user activates a location service with GPS turned on, the cell tower IDs are uploaded to their servers to help improve the Cell ID to coordinate mapping databases. On current mobile platforms such as Android, this is an opt-in service for users.

With these databases intact, phones could begin to easily report their location to any application that asked for it. Reverse-geocoding servers could then turn these long strings of latitude and longitude numbers into cities, streets, and addresses that could be used by a variety of applications. Early uses served as shortcuts for navigating to particular screens of a user interface that would otherwise require the user to enter a location by hand. Restaurant and movie search applications would automatically list relevant venues nearby instead of requiring a user to know where they were and enter an address or zip code.

Services like Loopt (2011), a California startup, emerged that allowed users to share their location with each other and view friends' locations on a map. In some of these services, users could leave messages at a location for friends to receive as a type of virtual graffiti to advertise their presence. As use of these features expanded, location sharing became a way to opportunistically meet up with friends who happened to be nearby, or to see friends who were bored and looking for something to do.

At the same time, other applications emerged to help people navigate. Google Maps released a version of their application that would center itself to the user's current cell tower. On phones with GPS, the application allowed users to follow a blue dot along the route to see when they were approaching turns. Later, full GPS navigation applications for mobile phones were released by TomTom and Google that allowed for full turn-by-turn directions similar to single-purpose GPS devices.

Now, social check-in services such as FourSquare, GoWalla, and Facebook Places are attracting millions of users to actively share their locations with others. While extensive studies on the use of these systems are just beginning, initial results of studies are demonstrating the potential of services like these to create opportunistic meet-ups as users see that friends and family are nearby, as well as to build an identity for others to see when using the service. Henriette Cramer from the Mobile Life Centre in Stockholm (Cramer, Rost, Holmquist 2011a) has completed the most extensive study of FourSquare use to date and has observed how these services create new urban games, motivate people to get out in the city, share their adventures with friends, and catalog places for their own later reminiscence.

Additional sensors worked to improve the accuracy of existing applications and provided new opportunities. Many mobile phones now come with digital compasses, which allow for simple improvements such as pointing a map in the proper direction when the phone is held out, but also allow for a new class of applications called Augmented Reality (Milgram and Kishino 1994).

Augmented Reality applications allow users to hold out their phone and see information displayed on top of a camera viewfinder. This makes it possible for virtual annotations to appear on top of physical objects in the world. For example, some applications allow users to see the nearest subway stations by holding out a phone and turning around. Station labels appear over the view in the direction of each station and the size represents how far away they are. These new applications enable a new class of context-dependent interactions that are very different from traditional fixed-location computing.

Accelerometers, initially used to determine the orientation of the device, began to be used for gesture input. The popular iPhone application Bump (2011) allows for users to share contact information, photos, and files with others by bumping their devices together. The system uses the accelerometer data to determine which devices are being bumped and then sets up a communication channel between those devices. Other applications are using physical gestures for gaming (e.g., turning a race car in a game) or on-screen navigation.

Another interesting area of mobile sensing, and one that will potentially impact our lives the most, is the use of body-worn sensors that communicate with the mobile device. Commercial systems such as Nike+ (2012) and MOTOACTV (2012) allow joggers and walkers to keep track of their workout distances and intensities both while they are running and afterwards on a dedicated website. The Nike+ sensor slides into a shoe and measures each step the runner takes. From this, it can determine speed and distance traveled, which are reported in real time on the phone display (and can be spoken aloud on demand and even shared in real time on social sites such as Facebook). After the workout, a graph of speed over distance is displayed for the user to analyze the intensity of their workout. Devices like this have the power to motivate people to stay in shape, as demonstrated by the research of Sunny Consolvo and colleagues at Intel Research Seattle (Consolvo et al. 2008). Her work with pedometers connected to mobile phones found that seeing a visualization of how far you had walked in a day and seeing feedback toward a goal helped to get people to walk further over time. This work used waist-worn pedometers that communicated to a mobile device over Bluetooth.

Other applications emerged that used location in a more social way. Jogging Over a Distance, developed by Mueller and O'Brien (2007), allowed friends in different cities to go for a jog "together." Both participants would don Bluetooth headsets and a GPS-enabled mobile device as they took off for their run at the same time. By tracking both runners' speed via GPS, the system knew when one runner was "getting ahead of" the other in terms of total distance traveled. As they ran, the audio of the person in the lead would appear to move to the front and get quieter, thus encouraging both runners to run at the same pace so that they could keep up their conversations while working out.

In 2008, Nokia began using location data to better predict travel times on freeways in California (Amin et al. 2008). Users agreed to upload their location during their commute to Nokia's servers, and Nokia could then obtain real-time information about conditions on roads all over the bay area and feed them back to users so that people could better plan their routes and departure times. This type of data is now standard in products such as Google Maps and data from users' devices is fed back to make the travel times and traffic conditions more accurate.

The tourist game REXplorer, developed and fielded by Rafael Ballagas and colleagues from RWTH Aachen and ETH Zurich (Ballagas et al. 2007), took visitors in Regensburg, Germany on a magical quest through the centuries-old city center. By using sensors such as GPS and accelerometers, visitors received information about city landmarks and were able to cast spells by moving their handsets in particular motions. Ghosts from the city's past would come alive and share their experiences with the visitors, taking them back in time to life in the city decades and centuries ago. These types of applications brought greater interactivity for visitors and helped to bring alive images of the past for people physically navigating through streets that were hundreds of years old.

Current research prototypes are adding even more complex sensors to mobile devices. Research at Carnegie Mellon University is adding CO_2 and particulate sensors to devices so that users can observe the pollution in their daily lives (Kim and Paulos 2010). A field trial in Africa led taxi drivers to change their routes as they became more aware of pollution problems in their cities.

Applications that take advantage of these rich context-aware sensors require detailed attention to the environments of use and testing in the field over longer time frames, which is where the methods in the remainder of this book will focus. How a user reacts while out on a run or seeing the actual pollution readings for their drive to work will be quite different from anything that can be observed in a traditional lab setting. Since it is impossible to observe a user directly over many weeks, existing methods of contextual inquiry and usability evaluation need to adapt to these more dynamic environments.

***

We have just explored how mobile devices have enabled new forms of communication and interaction with our world. Using a mobile phone today requires complex interactions involving context, media capture, and social connections. In order to create and evaluate these new types of experiences, new context-dependent methods are required.

The remainder of this book will be an exploration of some of the methods that we have adapted or created for building novel mobile experiences. These methods have evolved over years of application in Motorola's Applied Research Center and through a class at MIT called Communicating with Mobile Technology (Bentley and Barrett 2011) where we teach these methods in a project-based course.

We will begin by exploring the traditional user-centered design process, from which many of our methods derive. This process is seated in a rich tradition of observing users in their real context of use to innovate and test the effectiveness of solutions in their lives. We will explore how the unique aspects of mobile devices mentioned above, such as the inability to directly observe users in many of their daily uses of mobile applications, require rethinking some of these methods.

Chapter 3 will introduce methods for design inspiration. By grounding research and design teams in current behaviors in a domain of interest, we have been able to better anticipate user needs and expectations and have been inspired by the many ways that our participants have tried to solve their everyday problems and live their lives.

The next five chapters will focus on turning design ideas into functional prototypes, from design, usability, implementation, and validation perspectives. We will discuss methods to quickly prototype and validate key components of a proposed experience, including ways to get a working prototype into the field as quickly as possible to rapidly ensure that design hypotheses are true in a particular instantiation of a system.

We will conclude with a discussion of turning prototypes and designs into successful mobile products, including the multiple ecosystem partners involved in releasing a mobile application as well as some of the new methods of distribution, such as application stores and mobile web applications.

By sharing our experiences and methods, we hope that anyone involved in developing novel mobile applications and services can learn new approaches to better create experiences that fit into people's everyday lives and to take advantage of the unique properties of the modern mobile device.

2 The User-Centered Design Process Applied to Mobile

Context sensing. Media capture. Social connecting. The mobile experience is defined by these unique and fluidly integrated processes. It might sound simple in theory, but how does one get from a domain of interest to a new application or service in public use? This chapter presents an overview of a user-centered design process for creating powerfully useful mobile applications that can transform people's interactions with their friends, cities, and media. These methods will be developed more fully in each of the following chapters.

It is important, however, to differentiate "context, capture, and connection" in the mobile experience from how those states have come to be defined by our experiences with computational technology tethered to a desktop computer or a laptop. In the world of web design, "Context" often refers to users trying to locate themselves within, and navigate through, a virtual world of information. These tethered computing devices generally lack the ability to sense the world around them. Interacting with a laptop in an airport lounge is by and large the same experience as using it on a table at home. A mobile device adds rich contextual sensors and is aware of the world around it. Context is now much more than knowing where you are in an interface—it is where you are in a densely rich real-world environment. No longer is it enough to present a "map" of archived, published information. No longer is it enough to simulate a virtual world. The mobile device must be able to sense where a user is and facilitate actions situated in an immediate, living moment of experience defined by real places and times, by real states of being.

For example, in our class many student teams propose projects that help a person accomplish tasks within ordinary real spaces: in a car, on a bicycle, or in a living space. One team proposed building an application to expedite the drudgery of grocery shopping for a large family: how to navigate quickly and economically the intentionally confusing maze of supermarket aisles (the confusion helps distribute customers throughout the store in order to distract them with items they had not intended to, but might, purchase)—an especially helpful application as anyone who has done a week's worth of grocery shopping for friends or family can attest. The project team thought it would be a relatively simple task to create an application that contained your shopping list and in a variety of ways helped you navigate quickly and simply to those items. Their initial idea, however, was largely modeled after static desktop and web to-do list applications; it was not until they were instructed to go into the local supermarket and observe shoppers and get a feel for the layout of the store did they realize how limited their original conception of a solution was. They returned to visit this store many times, closely observing shoppers as they pushed their carts around to discover the intricacies in designing a useful mobile application for this situation. They found that shopping in the context of a store was much more complex than maintaining a simple list for an online grocery shopping service. They observed that people often saw particular items while shopping, and that this prompted entirely new choices of what to make for dinner and the selection of other items that needed to be purchased to make these dishes. Instead of being undesirable, this was a part of what made grocery shopping "fun" and should be supported in resulting applications, not avoided by forcing users to stick to a list and optimal route.

A powerfully useful mobile device allows a person to take advantage of the necessary and sufficient elements of that physically situated experience: location, time, visual and auditory characteristics, all that is apparent to our sensory apparatus—as well as data of physical states not immediately apprehensible. For example, another team in our class wanted to design an enhanced alarm clock application. They recognized that just setting an alarm was often inadequate if certain important elements of your initial waking decision had changed while you slept. Perhaps after a long week of exams you just wanted to sleep late on Saturday. But you were also an avid skier and your favorite New England ski area received an unexpected snowfall overnight. The team wanted to build an app that collected weather, ski conditions, traffic, and bus data overnight while you slept—and if certain parameters were detected your cell phone would automatically awaken you earlier (or let you sleep in) so you get to that fresh powder on the slopes (or your office door) at the optimal time.

And, finally, a powerfully useful application will connect you to other people to share facets of a contextually embedded experience—a real-time, instantaneous connection, or lagged, asynchronous interactions depending on the context. For example, the Motorola Contacts 3.0 solution for MotoBLUR (Bentley et al. 2010) allowed for an asynchronous feed of social network updates and photos from friends that could be viewed or responded to at the user's convenience. Or as one of our student's class projects proposed, a workout application that offered a range of instructional and archival tools for individual workouts but was also a tool for sharing your workout stats with coaches or team members or friends in another city as a way of finding new workouts from trusted sources while at the gym.

Process of User-Centered Design

The richly human, intricate, and diverse qualities of these three defining characteristics of mobile devices require an equally rich and subtle process for designing mobile applications. A user-centered design process emphasizes actual, real-world human experience as the primary and necessary inspiration for application design. Generally, it follows a recursive pattern from observation to design to implementation. Writing code to instantiate an idea (powerful as that idea may be) apart from an awareness of real-life interactions around a particular activity or set of activities may, at best, produce something that nearly addresses a person's needs; at worst, a code-centric design process results in a product that is usable only by its programmers.

These methods build upon a rich history of work in the Human Computer Interaction (HCI) field. This field encourages designers and product development teams to understand human needs and build new solutions to those unmet needs. Norman (1990, 1998; Norman and Draper 1986) spoke and wrote extensively on designing solutions that were easy to use and instantly understandable, like a doorknob or teakettle. Nielsen (1993, 2003, 2011) wrote extensively on the need to promote usability in design and include design from an early stage in product planning. Researchers Hugh Beyer and Karen Holtzblatt (1998) developed the Contextual Design method to better understand human needs and to inspire designs of new solutions while Paul Dourish (2006) and others have been promoting more ethnographically inspired ways to learn about interaction practices in the world for use in design.

The history of web design is filled with examples of the disregard of actual people with all their human capabilities and flaws. One example from an early project to design software for networked writing classrooms at MIT will serve as an instructive example. Talented programmers spent months of labor coding their own ideas into tools for a variety of hypothetically interesting quantitative analyses of written documents, including rather complicated tools for the automatic assessment of the quality of a written passage or document. The trouble was, none of these tools spoke to what writing instructors wanted to do with a richly networked communication system that was entirely text based. It took a minor coup by a diverse cadre of faculty and staff in the MIT Writing Program to place writers and their students in the design chair (Barrett and Paradis 1988), using writing itself to create a proper requirements document and

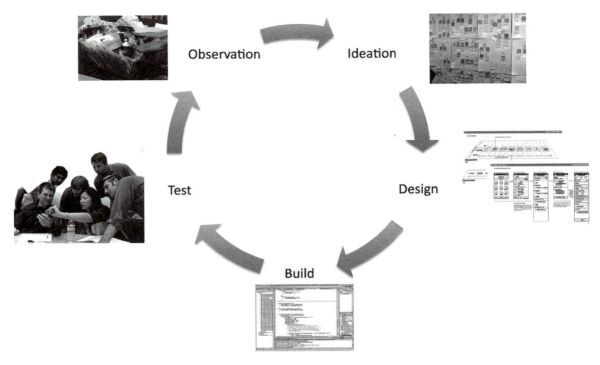

Figure 2.1

The mobile user-centered design process. The remainder of this book will discuss the methods that best apply to mobile design and development and how traditional design and implementation methods can best be adapted to the mobile environment. (Images used with permission of Motorola Mobility)

a prototype user interface built entirely around the processes that defined writing and teaching writing. Human experience became the design model rather than a theoretical demonstration of elegant coding processes.

In our class we apply that valuable lesson. Building the mobile experience is more than writing code. It is a process—halting, recursive, more meandering than linear—a process that builds from real-world observation and experience to ideation, prototyping, and testing.

Generative Research Methods

Discovering a meaningful new experience that you want to create for people is usually the hardest part of the process. Generally in our work, we start with an application space of interest. Both at Motorola and in our class at MIT, we seek to explore that space to uncover potential opportunities for novel applications and services that meet unmet needs, bring people closer to friends and family, or increase people's happiness and general wellbeing. At Motorola, we have explored music use, photo sharing, sports media, long-distance intergenerational communication, family coordination, social television viewing, and many other domains of interaction. Our students have explored grocery shopping, international travel, group coordination, meal planning, sleep patterns and alarm clock use, mobile game playing, and dozens of other domains. From these spaces of interest, we develop studies that answer key research questions and that contribute to our understanding of current practices and the types of experiences that are missing or are overly complex to create using today's tools. Many methods come together to contribute to this understanding. At this stage, we mostly use qualitative methods. Interviews, logging, photo and video diaries, home tours, and content analysis are methods we frequently employ to understand current practices. At times, we augment this with larger quantitative studies to see larger trends and group-based differences. We then use the data and analysis to inspire the design of new ideas for applications and services. By basing new concept ideas in data from a diverse set of participants, we are able to create new applications and services that work with the complexities of daily life and the interests of our future users. Investigating current practices within the context of generating new concept ideas will be the focus of the next chapter.

Design Methods

Good design for mobile services is critical to the ultimate usability of the final solution. Mobile devices have smaller screens and more limited space for interaction than a web page or PC application. Yet they also have many more input mechanisms: from the touchscreen and keyboard to location sensors, accelerometers, cameras, and more. Getting the interaction design right so that screens are not overly cluttered and users can flow from one task to the next in a manner that is easy to understand is a hard job to get right. Mobile design also needs to take into account everything going on around the device and how particular on-screen interactions fit into larger social or place-based interactions. Design for mobile is a very iterative process often involving many rounds of paper prototyping, usability trials, and the field evaluation of key concepts in order to understand how a given design works in many different situations of daily use.

Rapid Prototyping and Field Evaluation of New Concepts

Often, a generative research project will produce hundreds of design ideas for new applications or services in a given domain. Prioritizing these and quickly creating functional prototypes to test the new experience in the world is a key way that we can quickly identify which are likely to be successful in real-world use. As Nielsen (2001) says, "Pay attention to what users do, not only what they say. Self-reported claims are unreliable, as are user speculations about future behavior." Storyboards, focus groups, and asking people what they think about abstract concepts will not lead a design team to an understanding of whether a concept will be successful. To understand real behavior and what users actually would do with a new concept, we build a basic functional prototype in a matter of days or weeks in order to try out the new experience ourselves and explore it with others to see how it fits (or does not fit) into their patterns of daily life for weeks at a time. With little effort, we can decide to kill a project at this stage without investing too much time or effort. But if the trial proves successful, moving on to higher-fidelity prototypes and commercial products is often the next step.

Our MIT class operates in much the same way, but in a compressed time frame. In week one, students conduct a generative research project. By week two they have a concept inspired by that data. Week four brings the design of that concept and it is quickly implemented after that. Students evaluate their concept in terms of usability, but also in the way that it works in daily interaction. By the end of the term, they have a new concept, completely built, and have used it in their own lives. Many students then go on to launch their applications commercially in marketplaces for the various mobile platforms. Some students even go on to seek venture capital money to pursue their ideas further.

Commercial Deployments

When deploying a system commercially or on a large scale, there are many methods that can be employed to understand usage in the wild. Various strategies of instrumentation and targeted field studies can collect both quantitative and qualitative data that can be combined to understand how a system is being used in the lives of its users. Moving an application or service from the lab to the field is easier than ever with today's mobile platforms and application stores but still requires a good deal of thought and effort to help make a commercial offering successful.

For the past ten years, Frank has been involved in a series of projects at Motorola that largely have followed this process. The entire team in what has been called the Applications Research Lab, User Centered Solutions Lab, Experiences Research Lab, and then Core Research Lab worked together over years and numerous projects to refine many of the methods that appear in this book. Project team members will be recognized as each project is discussed.

The methods that will be discussed in this book are presented in table 2.1. These methods can often be used in multiple points of the research and development cycle as noted in the third column. For example, interviews and diaries are used in almost all of our studies in order to better understand details of interactions.

Before getting into the details of the methods, an overview of some specifics examples will help to demonstrate how these methods can combine over the course of a project. Table 2.2 illustrates several projects from our group in Motorola Research that have been taken from generative research through to products. In many of these cases, the ultimate product looked quite different from some of the original concepts generated, and the process led us through several iterative rounds of prototyping to new understandings about how these new types of systems work in daily life. This in turn led us to the ultimate solutions that we took to market.

We will refer to these examples throughout the book, so a brief introduction of some of the major projects will assist in putting each into context in later chapters. One of our broadest projects studied the area that we named "Ambient Communications." In 2005, we were interested in the ways in which people could receive social information and communication passively throughout the day on their mobile devices. This was a time before Facebook feeds and mobile photo sharing, and the only common way to keep up with friends was through explicit person-to-person phone calls and text messages. In the early stages of the project, we hypothesized that an awareness of the context of a close friend or family member throughout the day would bring about new ways of thinking about and interacting with that person. We thus began by exploring how people currently talked about location in their phone conversations (Bentley and Metcalf 2008), studying photo-sharing practices, and observing how people currently play, select, and share music (Bentley, Metcalf, and Harboe 2006). These studies inspired many concepts including Music Presence, where users could see the artist and song title of music that their friends were playing (Bentley and Metcalf 2009b); Motion Presence, where users could see if their close friends and family were at a location or in transit at a glance in their phone book (Bentley and Metcalf 2007); and Photo Presence, a system where users could view streams of photos from friends and family and comment on them (Bentley and Metcalf 2009b). These systems taught us quite a bit about the ways that streams of content coming in throughout the day can affect the way that people interact with each other and their environments. From there, we joined with a product team to create the Contacts 3.0 phonebook for MotoBLUR, a social network aggregator that allows users to see all communication and all updates from each of their friends at a glance on their home screen or in their phonebook entries for their friends (Bentley et al. 2010). MotoBLUR is currently

Table 2.1
Methods that will be discussed in this book for creating mobile experiences, with the stage of research in which they can be used as well as the benefits of choosing particular methods.

Code	Method	Stage	Benefits
HT	Home Tours/Field Visits	Generative, System Evaluation	Understand physical context of interaction, effects of environment on interactions.
D	Diary/Voicemail	Generative, System Evaluation	Receive details of interactions closer to the event so that more details are remembered.
CI	Contextual Inquiry	Generative, System Evaluation	Understand how people approach tasks in the context of their actual environments.
CL	Communication Logging/ Conversation Analysis	Generative, System Evaluation	Understand how users talk to each other about certain topics or details of interaction that are shared.
I	Semi-Structured Interviews	Generative, System Evaluation	Receive deeper information about use, answer "why" questions from more quantitative data collection.
CA	Content Analysis	Generative, System Evaluation	Understand the types of content that a given system affords.
FT	Field Trial of Prototype	Generative, System Evaluation	Understand use of a system in real-life situations.
IN	Instrumentation	System Evaluation, Commercial Deployment	Quantitatively understand use of system over time. Optimize flow for common tasks. Plan for scalability. Find specific instances to discuss in qualitative interviews.
CM	Concept Model	Design	Creating a model of the system that you are creating from the point of view of the user.
IM	Interaction Model	Design	Creating a model of the flow of an application to better understand how pieces will fit together.
PP	Paper Prototyping	Design	Quickly iterating on a design to explore different interaction models and flows.
UE	Usability Evaluation	Pre-System Evaluation	Ensure that users can understand prompts and icons, flow of system.

Table 2.2

Projects from the Motorola Applied Research team that will be discussed in this book and the methods that were used in each project. Methods are in parentheses and are explained in the previous table.

Project	Generative Research	Concepts	Product(s)
Ambient Communication	Location Sharing (CL/CA/I)	Motion Presence (I/D/FT/IN/P)	Contacts 3.0/ MotoBLUR
	Photo Sharing (I/HT/P)	Photo Presence (I/D/FT/IN/P)	
	Music Context (I/HT/CI)	Music Presence (I/D/FT/IN/P)	Connected Music Player
Sports Media	Observations at Football Match (I/HT)	TuVista Phase I (I/FT/IN/P)	TuVista Phase II
Media Metadata	Photo Sharing (I/HT/P)	Location-Based Organization (I/P)	J2ME version of ZoneTag
	Music Context (I/HT/CI)	Media Assistant (I/P)	Media Finder
Intergenerational Communication	Elder Communication (I/D/HT)	Serendipitous Family Stories (I/D/FT/IN/P)	StoryPlace.me

shipping on millions of Motorola phones and is allowing people to connect with the happenings of their friends and family at a glance throughout the day.

As you can see, this process was a winding journey to understand how people can best interact with streamed social data. We had many questions to answer about the type of data, the people the data should come from, the scope of who could see a user's data, and the interactions that should be supported from the data itself. On top of these technical questions, we wanted to know how these concepts affected relationships and interactions among family and friends and how we could make people feel a stronger sense of "connection" to each other over a distance. Through studies and prototypes, we answered these questions in small teams that could operate rapidly. In the end, our solution was able to connect people in new ways and successfully integrate into the daily patterns of life, creating rich connections across large distances in ways not possible with other technologies of the time.

Other projects at Motorola have traveled on a more linear trajectory. In 2008, we became interested in mobile sports video. After a bit of research observing the use of media at sports stadiums and some existing solutions in the mobile sports media domain, we created a rapid prototype of the TuVista system. We tested this system at a live semi-final game of the South American Cup soccer league. We learned a lot about the use of the system, which led to changing our business plan a bit as well as some changes in the design of the system itself. But in this case, the final product version of TuVista ended up looking and interacting very much like the early prototype. Bundles of media content were delivered to mobile phones shortly after plays occurred. Fans were able to watch video on-demand and learn more about the players involved in a given play (Bentley and Groble 2009). In this case, the process helped us to further refine the interaction and technical architecture as we moved from research to product.

Another project that has followed a more linear path has been focused on the topic of intergenerational communication. In 2009, we became interested in the ways in which we could use mobile devices to help people maintain stronger relationships with their parents and children over a distance. Populations in America and Europe are getting older and people are increasingly moving away from family. We wanted to investigate how families kept strong social connections over a distance and what we could create to make them even stronger. We began with a study of current communication practices, focusing on older adults in Florida and their adult children in Chicago (Bentley, Harboe, and Kaushik 2010). From this study, we observed how storytelling brought families together and helped to reinforce their identity. We also observed an instance in which a participant was walking near a theater in Chicago and started talking to her mom about her grandfather who used to dance there. We were inspired by the connection between place and story, and went on to create the Serendipitous Family Stories system (Bentley, Basapur, and Chowdhury 2011). The system allowed users to create video stories and save them at a place in the world. They then could share them with friends and family who would receive notifications on their phone when they happened to walk near a story location. The recipient could then view the video in the context of where the story occurred. This connected participants not only to their friends and family, but also to their city in unexpected ways. We saw the power of this contextual storytelling to change people's awareness of city and family history and wanted to release it as not just an intergenerational storytelling tool, but as a system that anyone could use to connect to their friends or city. The new service we developed also included professionally produced historical content about places in the city that could be discovered in a similar way. That resulting system, StoryPlace.me, is now live on the web and in the Android Market (Bentley and Basapur 2012).

Evaluating the Results of All Methods

Ultimately, these methods must be evaluated by the end results. These can be negative results as well as positive ones. Clearly, inventing new applications and services that are widely adopted and transform the lives of their users is a clear win. But likewise, being able to stop development on ideas that do not show promise or redirect efforts in ways that are connecting with users is also a success. Many of our concepts do not end up looking anything like what we had in mind when we started. If used properly, the methods in this book can help researchers and designers hone in on the high-value and high-impact mobile applications and services while identifying failures early in the process.

3 Discovering What to Build

Discovering what to build is often the hardest part of creating new products that seek to fundamentally change the ways in which people interact and communicate. By ensuring that a new concept is rooted in users' behaviors and needs, often actually designing and building the resulting system turns out to be quite simple. Getting the right concept, however, is often difficult and the result of much trial and error. The goal of good generative research is to increase the percentage of hits that get taken forward to successful solutions that can help people in their daily lives. There are many ways to be inspired, but we have found that engaging with people out in the real world and learning from their experiences is the best way to get that spark of an idea that leads to an invention.

Following the general pattern of user research outlined in the previous chapter requires some modifications for the mobile environment. Mobile interaction takes place in the contexts of daily life, often in quick bursts and in parallel with other tasks such as waiting for a train, navigating a city, or collaboratively planning activities with friends and family. We have found that observing these interactions between people, media, and context can often be the best inspiration for new concepts, as we can find the key tasks and behaviors that are currently unsupported by today's mobile applications and services. We then use insights from our observations to help in designing and building that new experience. This requires getting data from a variety of places and situations outside of the lab environment and engaging with users to understand their lives. It also means using creative methods to "observe" users' interactions at times when a research team cannot be looking over their shoulders.

Paul Dourish from UC Irvine (2006) saw the way that direct observations in the world can create ideas for new products and services and called these findings from research most valuable as "inspiration" for design in a much-discussed 2006 paper at the CHI conference on Human Computer Interaction. These inspirations differ from often dry and detail-free design guidelines that historically came from field research in design. By directly sharing observations of users and quotes from interviews, researchers and designers can be inspired to build off of subtleties from field observations in creating new concepts. We wholeheartedly agree with this approach and view our early-stage ethnographically inspired work as a means to get smart in a design space and to get the creative design concept ideas flowing.

When starting to work in a new space (or even an old one after some time has passed and new practices have evolved), it is important to understand how people currently interact with technology and each other in the domain you are interested in. Understanding what does and does not work for them is often the best way to invent new concepts to help with unmet needs or to make sure your design ends up fitting into the ways in which people want to interact with each other and with the content provided by your system. Participating in this type of research not only prepares a team to design new concepts in that space but turns them into domain experts, or at least advanced novices, to be consulted on further projects in that space. This knowledge often stays relevant for some time after the studies are complete, as they are meant to create an understanding of deep human behaviors and interactions that are less dependent on the current state of technology. We have applied these tools in many domains over the years before creating designs and prototypes in these new areas.

In 2006, Frank and Crysta Metcalf became interested in location-based services and the concept of sharing one's location with close friends and family. However, we saw many problems with existing location awareness tools that shared absolute location at all times and seemed to amount more to tracking than facilitating healthy social interactions around locations. We hoped that we could create a system that would fit in better with people's true uses of location sharing in their current social interactions and would enable people to engage in new opportunistic interactions with the new information. We wanted to start by looking into the ways that people used location in their daily lives in order to provide some inspiration for the design of new concepts. By understanding the reasons for sharing location in everyday phone calls, we hoped to better understand how location sharing fit into patterns of everyday interaction out in the world. With this knowledge, we would be in a better position to invent new services that fit with the nuances of everyday life and were better aligned to people's real-world privacy concerns. We believed that this understanding would be critical in exploring why existing location-based services were failing to work in people's lives and would lead to insights to allow us to create new location sharing concepts that fit with people's motivations for sharing their location with others.

We devised a study in which we gave a set of diverse participants phones that would allow them to record their phone calls (with permission of all parties) for a week and save them automatically as time-stamped files on a memory card in the phone (Bentley and Metcalf 2008). Midweek and at the end of the week, each participant would FedEx us the memory card from their phones so that we could transcribe the phone calls and interview them about their location sharing. We were interested in the ways in which they talked about location with each other, the purposes for this sharing, and their effects.

From this study, we learned that people in strong-tie social relationships often share the details of their current location and near future plans with each other over the phone and in person. This sharing allows close friends and family to have a fairly accurate mental model of where their friends and family are at most times of the day. However, what is often unknown is the transition time between these locations. Participants in our study frequently asked questions to confirm their exiting mental models such as "Have you left yet?" or "Are you home already?" We also saw that being unclear about someone's location sometimes led to no communication taking place at all, for fear of interrupting someone when they were at work or driving.

With this finding, we created a concept, Motion Presence, which helped people to know when their close friends and family were transitioning between places, without sharing the actual details of their locations with each other (Bentley and Metcalf 2007). In this concept, an augmented phone book showed if particular contacts were "at a place" or "moving" between places and for how many minutes they had been in that state. However, it did not share any details about what that place might be. In later field trials, we saw that participants were able to get an understanding of availability from the data provided and used this information to decide when to interrupt their friends and family with a phone call. They also appropriated this information for all sorts of unexpected uses including getting more time at their current activities if they saw that someone was running late or using the data to know that a loved one was safe and moving about as they usually do throughout a night shift or commute home late at night. The insights from this concept eventually led us to the presence-enabled phone book in MotoBLUR which is now used by millions of people to stay up to date with their friends and family and to coordinate activities.

Another idea came after mobile devices began to add significant amounts of user storage space in 2004. Devices with Secure Digital (SD) cards were on the horizon and we saw the inevitable future where phones and MP3 players combined into single devices. Because of this, we became quite interested in how people interacted with their music collections. Frank, along with Crysta Metcalf, Gunnar Harboe, and Vivek Thakkar at Motorola Labs developed a study to investigate how people selected music to play and how music was stored and acquired (Bentley, Metcalf, and Harboe 2006).

We knew that interacting with large lists of content on mobile devices was going to be difficult, as was already seen with the scroll wheel on Apple's first iPod. We knew that there had to be a better way to select the right music for a particular context and went out to study this phenomenon in daily life.

In the study, we visited people's homes and learned how they organized music, selected music for particular events, and how they acquired new music. We learned that music selection was often an iterative process whereby a user would pick something to play and then desire something a bit different—a little faster or a little slower, newer or older, something they had not heard in a while, or their latest acquisition. We also saw the desire for the ability to seek out "more like this" in particular ways: perhaps more from that artist, more in that genre, or more from that time period. We observed that this process of music selection follows a satisficing behavior where people would stop when they reached some music that generally fit into what they felt like listening to at a given time.

From these observations, we created two concepts. The first was the Metadata Knob, a dial that allowed a person to tune her music collection just like a radio (Bentley et al. 2008). Just set the knob to beats per minute and dial in the music for your current mood. As you turn the knob, you hear different music start to play, and you turn until you hear something that you like—just like a radio. Less than a week of development later, we had our first prototype. Our second concept was a play tree, where instead of a linear playlist that goes from one song to the next in a predefined order, users could see a list of next possible songs based on different metadata attributes of the currently playing song (Harboe et al. 2008). One option might be another song from the same album, one might be from another song in your collection that was released that same year, and another might be one with the same tempo. If the user did nothing, the next song in the list would be selected automatically. But if they wanted to be a bit more interactive, clicking on their choice for the next song was an easy way to help direct the music playing to be more like what they were looking for.

These examples are just a few of many from our work at Motorola. In our class at MIT, our students follow a shortened version of this process in the first week of the class to help seed their minds with data from real user behavior in the areas in which they are interested in developing applications. Some projects have investigated supermarket shopping, overseas travel planning, couch surfing, or alarm clocks. By getting a quick understanding of how people currently perform these activities, students are able to ground their design work in more than just their own experiences and create better overall systems. This process does not need to be long (we will discuss ways to shorten it at the end of the chapter), but it is valuable before designing or building technology in a new space. You will often be surprised how your assumptions differ from reality and how much Lund's (1997) mantra—"you are not [like] the user" of your intended system—holds true.

The Process of Generative Research

How does one get from a space of interest (e.g., music, location sharing, intergenerational communication, etc.) to a list of potential solutions? The work of Crysta Metcalf and the entire team in our lab at Motorola have refined this process over the past decade. Together, we have tried many different methods for understanding current behaviors in an area and applying this knowledge to create design guidelines and ideas for new concepts. This process takes the researcher on a path from well-defined research questions to methods, data collection, analysis, and ideation. A study that is rigorous and publishable might take a few months of work, but discount methods allow for much faster cycles and can take as little as a day or two. Depending on the goals of the team, this stage in the process can take an amount of time anywhere in between those extremes.

Defining Research Questions

The first step in any research process is to formalize the questions that need to be answered. These are generally quite broad questions about your domain of interest and will lead your team to specific insights on user behavior. Because these studies tend to be small and qualitative, these questions should be more about understanding the breadth of use (the how and why) instead of the quantities of use (how much and for how long). Since the rest of the study and the data you collect will be based on these questions, it is important to give some thought to all aspects of your domain that you might find interesting. It is always easy to prune back the list of research questions when designing interviews or probes, but this first list should be as all-inclusive as you can get. Lindlof and Taylor (2002) also include a detailed discussion on defining research questions for qualitative communications research.

The use of mobile systems is often contextual, and our research questions have tended to focus on understanding these contexts of use. They often focus on where and when given interactions take place and both the trigger that started the interaction and any follow-up action that was taken because of specific interactions. We are also usually quite interested in the social interactions between people. Sometimes this involves phone calls or messages between people but can also include interactions with people who are present, such as collaboratively selecting music or deciding on a restaurant in a group while mobile.

A few research questions from some of our studies can be used as a guide:

Intergenerational Communication Over Distance (Bentley, Harboe, and Kaushik 2010)

- How do participants communicate with remote family/closer relations? What tools do they use?
- What are the barriers to remote communication? What challenges do people face in maintaining or engaging in remote communication?
- What communication tensions and obligations exist surrounding remote communicate for the elderly?
- What are differences between communication at a distance and communication when remote relatives visit?
- What artifacts in the home serve to promote remembrance of and communication with distant family?

Use of Location in Phone Calls (Bentley and Metcalf 2008)

- During communications with others, what location and activity information is provided?
- Under what circumstances is this activity and location information disclosed?
- Why do people disclose location and activity information?
- How are disclosures of activity and location similar? Different?

Music Usage (Bentley, Metcalf, and Harboe 2006)

- What breakdowns exist in today's music experience in both independent and co-located situations?
- What contextual (from the past and the present) metadata can be used to address the breakdowns in today's music experience as identified in previous question?
- How can this contextual information be used to enhance today's music experience?

As you can see, these research questions are broad and aim at uncovering new insights about user behavior in the target domain. By keeping the questions broad, there is a greater chance that the research team will uncover unexpected findings that will lead to novel design ideas. These questions rely heavily on the context of the user and interactions in particular spaces and situations. With mobile services, integrating new experiences into the contexts of daily life is often the hardest part to get right.

From the research questions, a study must be devised that will answer the questions and lead to new design insights. There are many methods that can be appropriate depending on the types of questions that have been identified. Many of these methods serve to uncover interactions that are not directly observable by research teams or to further understand particular contexts of use. In the next few sections, we will explore some of these methods with examples of how they have been used in work throughout the research community and in our class at MIT.

Using Probes

Mobile behavior is difficult to observe. The most interesting uses often occur at times or places that are not directly observable by a research team. Often much use is contextual or emotional and cannot be replicated in a lab or home interview. Understanding these contextual uses is critical to understanding how today's technologies are used and the situations that lead to joy or frustration. As Grimes and Harper (2008) discuss, designing from observations of joy and celebration can create new concepts that are fun to use and build off of the best of our interactions with the world. However, since mobile interactions are often short and particular details of a mobile interaction are often unmemorable, simply asking people to recall particular instances of a behavior in an interview will not yield the detailed (or even accurate) data that is needed for design. And it will rarely recreate the emotions of the instant that are often critical in interpreting an interaction. This is where probes come in.

In short, a probe is a piece of technology that is given to a user in order to help researchers observe parts of their life that are otherwise hidden from interviews or direct observation. The original cultural probes, as deployed by Gaver, Dunne, and Pacenti in 1999, consisted of a camera, postcards, and a map. Frequently, studies using these methods are called "Diary Studies" as participants often keep some sort of journal of their interactions and environments throughout the course of the study. The details that are logged more closely to an interaction are more likely to be accurate than a recollection some days or weeks after the event. In this way, researchers can get firsthand accounts of behavior as close to the time of interaction as possible. The overall goal of a probe is to be as simple as possible for the participant to record in situ, while still providing meaningful data to the research team.

We can classify probes into several buckets. First, there is a set of probes that are meant to simply serve as memory aides. These probes often take the form of a log or voicemail diary. Whenever a participant does a particular action, they are asked to log it. Usually in the log, they are asked to answer specific questions that are of interest to the research team, such as the time and context of the interaction as well as specific questions that follow from the overall research questions. For example, in a study on intergenerational communication over a distance, we asked our participants to call into a voicemail right after any communication they had with their parents/children to tell us about the details of that interaction and what was shared while it was still fresh in their minds. Sometimes, these probes can be automated. Many mobile platforms now allow for monitoring of call state or messaging history, and automatic reminders with links to call into a voicemail or complete a survey can be presented immediately following the activity of interest to help increase response frequency while memories are still fresh. With these probes, we have found that keeping a regular frequency of calling (e.g., nightly) can help improve the likelihood that participants will provide data and not forget to call. Frequent calling also allows research team members to know when a participant has not provided information (even if all that they report is that they did not do anything you were interested in that day, the participant should still call in daily) and to remind the participant to keep calling in or completing surveys.

Another type of probe is meant to add visual information for the researchers that is not present in a verbal interview after the fact. These visual probes allow researchers to see into the lives and interactions that their participants experience. Often, these probes include photos or videos taken at the particular times and places of interaction. This can help researchers to better understand the context of an interaction and other forces that might be at play surrounding the user. For example, are there other people around that might have influenced the actions taken, are there other devices besides the phone involved in a particular interaction such as a paper phone book or computer, are there particular places that are frequently used for a type of interaction? Seeing the context can provide the types of insights usually gleaned from direct observation of users without the need to follow them day and night. These probes follow directly from the types of probes that Gaver's team used and help researchers to see into the lives of their participants and the sometimes unexpected contexts of use.

A final category of probes involves giving participants a piece of technology to see how they would react in a given situation. This technology is usually created to help answer a specific research question that might be impossible to answer without it. Consolvo and Walker (2003) at Intel Research applied the Experience Sampling Method (ESM) to mobile technology. In an ESM study, questions for a user pop up on a mobile device throughout the course of the day. In one study, the team at Intel asked users to indicate what fidelity of location (e.g., GPS coordinates, neighborhood name, city, etc.) they would share with a particular contact at the moment of interruption as a way to better understand privacy concerns around location sharing.

In a study in our lab at Motorola, Harboe and team (Harboe et al. 2006) placed a speakerphone between the homes of two friends who were watching the same television program in order to explore how people would converse over a distance while watching the same content. These types of probes are quite useful when studying phenomena that are not yet mainstream and in seeing how people would respond in these new situations. However, it is important to take the novelty effect into consideration and investigate how people continue to respond to these systems over repeated use. Insights from using a technology probe may be more accurate than an interview or focus group, since participants have actually used the concept; still, it is important to recognize that this use remains an initial impression and is not representative of use over time.

Logging and Conversation Analysis

Since mobile interactions are often quick and take place in parallel with other activities, the details of these interactions are often not memorable. In many cases, the data that researchers need to inspire new concepts relies on specific details of how people are interacting with their content and with each other. Data such as particular lines of conversation or specific steps taken in an application are quickly forgotten, even if participants are required to create a diary entry or voicemail directly after their interaction is complete. In these situations, automated logging or conversation analysis can help researchers to get at the missing data and complete the picture of the user's behavior.

Many mobile platforms allow for recording phone calls directly from a service running on the device or with a simple user interaction. When both parties have agreed that recording of phone calls is acceptable (policies vary by state and country), having access to this information can be a valuable added source of understanding for how users are planning, navigating, and sharing information with others. Alexandra Weilenmann of the Viktoria Institute pioneered these techniques back before audio recording on mobile phones was simple. With portable tape recorders, she had users record their conversations over several days and analyzed the data to understand how people talked about location with each other. She and Peter Leuchovius (2004) observed the use of personally meaningful location descriptors such as "where we met last time," which could be used in designing systems for micro-coordination: the iterative task of planning, and then actually meeting with other people. These types of in-group location descriptions could also be used to mitigate privacy concerns as third parties are unlikely to be able to interpret this type of location data. In another study on availability, Weilenmann (2003) observed the use of location words in statements such as "I can't talk right now, I'm in a fitting room." This information was used by many researchers and mobile application designers to inspire various location sharing and mobile awareness systems that could help moderate availability, and it helped to tie together activity and location as frequently interchangeable descriptions of one's availability. These lines of conversation could likely not have been obtained in any other way, as study participants would probably only remember the vague details of their conversations if asked about them later. Here, the particular details in real-time speech provide the interesting data for analysis and new concept ideation.

In our own research on mobile location sharing mentioned at the start of the chapter, Frank and Crysta Metcalf (2007) at Motorola had participants record their phone calls with consenting friends and family and also analyzed location disclosure practices. We observed that people usually knew the basic mobility patterns of close friends and family, but often asked about the other party's location or shared their own to confirm transitions such as getting to work or being on their way to meet in person. This helped inspire many of our location concepts including Motion Presence (discussed in chapter 4) and the Contacts 3.0 service, which became the phone book in MotoBLUR in its commercial release.

Existing repositories of content can also be mined to understand user behavior. Researchers from corporations and academia have analyzed content on Flickr, YouTube, and Twitter to understand how people share content and build relationships online. Oskar Julhin, Arvid Engström, and Erika Reponen (2010) from the MobileLife Centre in Stockholm and Nokia Research analyzed the content of videos uploaded to mobile video sharing sites such as Bambuser and Qik in order to discover the types of media that are shared in these systems. They observed the relatively poor production quality of the videos, including transitions and poorly produced beginnings and endings. Their design recommendations include ways for mobile video broadcasting systems to improve these aspects of production and to tailor interaction with these systems to the types of content that are created while mobile, which they observed to be different from video content created while seated at a computer. The resulting systems would lead not only to better content on the site, but to users who are now better educated in the art of film production.

When building your own systems, building in the appropriate instrumentation so that you and others can analyze use or perform content analysis is important so that everyone can benefit from understanding how technology is being used in the world. The Facebook Data Team (2010) often releases data derived from their own content analysis, such as a recent exploration of voting practices in the 2010 midterm elections in the United States. For example, they were able to predict sixteen of the twenty close races just based on counting Facebook "likes" for various candidates. Similar content analysis can be performed on many existing systems. More information on instrumentation can be found in chapter 9.

Logging interactions and communication can help researchers to better understand current behavior and use these findings to inspire new solutions that are tied to real users' lives. However, this raw data can be open to many interpretations. All that can really be learned from a log file is that a particular phenomenon occurred. No explanation of this action can reliably be given without more qualitative data. Explanations of these practices can develop from combining the log data with follow-up user interviews, leading to deeper insight that goes beyond the raw data and provides useful insights into why certain patterns of use occurred.

Home Tours/Field Visits

A person's home says a lot about who they are and what they consider important. The organization of items in the home can also say a great deal about how participants use various types of media. Data from a home tour can serve to put data from other methods in context, showing the places where interaction occurs as well as the objects and information that are present in that environment. "Home" tours can also include other environments of interaction such as workplaces or transit routes to better understand the environments in which users interact with their devices and with other people or content. When studying topic areas as diverse as photo use, music use, television, or intergenerational communication, we have employed the home tour as a part of our study design to learn more about how these aspects fit into our participants' lives. Students in our class have performed field visits in grocery stores, dorm rooms, coffee shops, and many other locations around Boston to better understand the context of various activities from grocery shopping to travel planning and mobile game playing.

In our intergenerational study exploring practices of communication across distance and generation, we started with a tour of each participant's home so that we could understand the variety of places and devices that they used to communicate (Bentley, Harboe, and Kaushik 2010). As we went through their homes, we asked about the last times that they communicated in each room or on a given device. We were easily able to learn about the places from which they liked to communicate and the factors that led to them choosing a particular communication device or location. We were able to observe how some locations created spaces for quiet, comfortable

Figure 3.1
Photos from the home tours in the intergenerational communication study. Through our home tours, we discovered the importance of place in communication and the reduced use of communications technologies that resided in out of the way places such as extra bedrooms or basements. This led us to thinking more about the importance of place in communication and ultimately, in combinations with other findings, to the concept for Serendipitous Family Stories.

communication while others led to frustration or distraction. After the initial interview, participants called into a voicemail nightly to discuss their communications. Because we had visited their homes, we could visualize the places from which they reported making phone calls and the artifacts in their homes that led them to think about their relatives at a distance. Without the home tour, much of this information would not have been understandable from a static voicemail. We were also able to learn the diverse ways in which our participants integrated communication into their lives. When asked to show us where he communicates, one participant asked us, "You mean where the phone hangs on the wall?," while another participant talked about the wide variety of places both in and out of the home she communicates from including trains and tanning beds. Understanding these perspectives and visiting the environments where communication occurs helped us to better understand how place plays a role in communication, especially for older adults, and led us to focus on place-based content creation in our resulting Serendipitous Family Stories system (Bentley, Basapur, and Chowdhury 2011). In this system, users could leave video messages for friends and family in particular real-world locations. When recipients came near one of these locations, they would be automatically notified that a story existed there and would be able to view the video. This became a way to share family history in the rich context of the world in which it occurred.

In our music study from 2004, we also utilized home tours as a method (Bentley, Metcalf, and Harboe 2006). We were interested in the places in the home and car where

Figure 3.2

Photos from home tours in the music study. We observed the piles of physical music that our participants kept in various locations in their homes, cars, and gym bags, thus creating music zones where only particular music was accessible (left). We also observed the importance of gifted music in its original gifted packaging (center) and the ways in which our participants prominently displayed their favorite music in the home (right).

participants stored music as well as how they selected and played music in particular contexts. Most participants in this study still had fairly large CD collections as digital music and the Apple iPod had not reached a large market by this time. We could see how the music at hand led to particular music choices in certain contexts and how favorite or recently acquired music was kept at hand and regularly moved between car and home. We saw participants who got stuck in music listening ruts in their cars because of the limited set of CDs that were there as well as their strategies for getting out of these ruts and into new music or old favorites by rotating their collections. All of this information helped us to understand the physicality of music and some of the positive aspects of that physical nature that might be exploited in digital music systems. It led to particular insights in maintaining easy access to recently purchased music and to inventions like the Play Tree or Metadata Knob that could let users branch out from their currently playing music into other parts of their collections.

When conducting a home tour, it is important to focus on areas that directly answer your research questions. Often, these questions center on the places where a particular type of interaction occurs. What distractions might be present in that environment? Why is one area of the home used for this activity over others? What other items are placed near your items of interest? Through careful observation and non-leading questions that speak to the research objectives, researchers can understand how the environment impacts communication and content consumption behaviors.

These types of tours are often best conducted during an initial interview before the official data collection takes place. By seeing the home, the research team can build a deeper understanding of their participants' lives. This understanding helps to place in context statements made by participants during the later data collection phase of the study.

Home tours are also best conducted with a pair of researchers. We often have one researcher videotape the tour while another takes the lead in asking questions and following up on statements made by the participant. We often alternate the interviewer and videographer roles between researchers and the videographer is always free to follow up on a question or ask something that the primary interviewer may have overlooked. Depending on the depth of the interview, home tours usually last about half an hour and are an easy way to quickly learn about a participant's environment.

Task Analysis

While many mobile interactions occur naturally in the world, at times it is useful to create a contrived situation in order to observe how someone interacts with a particular piece of technology. To do this, researches often perform a task analysis. Beyer and Holtzblatt (1998) explored this topic in their paper and book on contextual design. They commented on the power of watching people perform tasks in the contexts of daily use to help researchers better understand their work processes and how information and materials flow while accomplishing a task. We have applied methods derived from this work in attempting to understand people's interactions with music and existing mobile applications.

In our study on music use at Motorola, in addition to home tours and interviews, we asked participants to perform some directed tasks for us (Bentley, Metcalf, and Harboe 2006). In an initial interview we had asked them about recent episodes of music search and the scenarios in which they have looked for particular music in the past. Then later on, while in their homes, we asked them to act as if they were in similar situations and to select and play music for these particular scenarios. For example, one user was asked to pick music to play while studying. Another was asked to pick music to play for a card night with friends. These were tasks that they reported normally undertaking at home, so we were able to see how they used the music piles that they had around them and the devices available for playback to select and play what they thought was the appropriate music for each particular scenario.

During a task analysis, it is common to have participants "think aloud" and verbalize what is going through their head at any given time (as described in Lewis 1982). This helps researchers to understand their thought process and why they are choosing to do the task in a particular way. The think aloud method is another way to understand what a user is thinking at various points of an interaction and not just what they are doing. By thinking aloud in real time, users are more likely to be stating what is actually going through their minds, rather than making up explanations if they are asked later why they acted in a particular way. Often, it is a bit uncomfortable for users to constantly verbalize what is in their head. Proper warm-up exercises, like having participants think aloud while trying to reload a stapler or setting an alarm clock, can help in warming up participants to speak while performing a task. Often, researchers will have to remind the participant to continue to speak while performing tasks. It is important to hear participants' thought processes as they go through each step in order to uncover usability issues and places where existing systems do not meet the intent of the user, such as when our music study participants wanted music that was a little slower or faster than what they were currently playing.

Task analysis can also be conducted out of context. In final interviews, we sometimes ask participants to show us how they have been using a particular system to help us clarify questions we might have from a voicemail log or interview. Seeing a participant interact in front of you, even if it is a task that she would normally undertake while mobile, can help you to understand the details of an interaction that she might not otherwise discuss in an interview or log.

From this analysis a number of models can be made. Beyer and Holtzblatt (1998) introduce several in their book *Contextual Design* that are quite useful, including diagrams of how information or content flows between users or places. These diagrams can be used to better understand how particular people or places encourage specific actions. For example, in our study of music use in the home, we saw how music tended to flow to specific places and "stick" there—for example, CDs brought into a car or bedroom. These findings helped us in brainstorming ways to get people out of their music "ruts" and on to different collections of music when they wanted something new.

Semi-structured Interviews

While other methods help to uncover details of use or interactions within a particular context, interviews still serve as a primary means to make sense of data that has been collected and to learn about how data collected during the period of the study compares to more typical patterns of behavior. All of our studies have included some type of semi-structured interview to better understand our participants' behavior.

Most of our interviews tend to be semi-structured so that we can explore areas of interest that our participants bring up in response to our discussion. These methods are derived from years of interview protocol design in fields such as anthropology.

The references for Bernard (2002), Angrosino (2002), and Ervin (2000) can help in conducting interviews and properly formulating non-leading questions.

We tend to focus our interviews on particular instances of interaction in our domains of interest (music use, phone calls with a particular person, etc.), usually starting with the last time our participants have interacted in a way in which we are interested. We then ask follow-up questions to better understand the context of that interaction and what happened next. As demonstrated in years of anthropological research, asking about the last few interactions can better help us to understand frequency of use and to have something to check against the data we collect during the rest of the study. This allows us to see if the data we collect during the study is typical of interactions that were performed before it without relying on people's poor abilities to recollect frequency data.

In final interviews, we often ask users to clarify data that we collected with other means, such as through a diary study or through logs from a mobile phone that they have been using. We use the existing data as a probe to help users remember the given context of interaction. The more data we have logged about the interaction, the easier it is to help users remember that particular instance. Reminding them of the day, day of week, time, and people involved in the interaction can help. Where additional data, such as what was discussed, is present, these details can also help jog a participant's memory. Showing participants a photo that they took and asking for more details around why, when, and where it was taken will lead to much more detailed and accurate results than simply asking participants to remember "the last time they took a photo." Rich visual cues from captured media and other log data can be the bridge to a memory that will result in a successful interview and add key missing explicatory data to the analysis.

Recruiting Users

For most generative research, we try to recruit between seven and twelve participants for each study. We have found that this number allows us to see a wide variety of use, but when using more than this number we begin to see a majority of repeated data with extra participants, which does not broaden our understanding of practices in the domain of interest. In order to achieve the broadest data set possible, we actively recruit participants that are as different from each other as possible. This generally means different ages, occupations, income brackets, genders, ethnic backgrounds, and technology use. For studies that examine a particular topic, we try to recruit people who have a wide range of use for that particular topic. For example, for the study in intergenerational distance communication, we recruited some participants who spoke multiple times a day and other participants who emailed or talked only once every few months. Finding patterns across participants who are very different often points to more fundamental human needs that will lead to more universal application and service ideas that can work in many peoples' lives.

Recruiting is one of the most important steps in executing a generative research project. The data that is collected and the findings are only as good as the participants that are selected for the study. In the end, the data is only valid for the participants you interview and observe. If all participants are male computer science college students who are eighteen to twenty-one, it is hard to generalize the results or uncover a wide range of behaviors.

It is often difficult to find this perfect mix of participants on your own. Emails to extended networks, posts on mailing lists or Facebook pages, and putting up posters at grocery stores or college campuses can only go so far in attracting a representative group of users. Often, we need to use a professional recruiting agency to find our participants, especially when there are geographical or phone-related constraints (e.g., participants with Android phones living in Chicago with older relatives in Florida). While a recruiting firm can cost more money than trying to recruit on your own, the resulting diverse participants that they can find are almost always worth the expense in terms of the richness of data that can be collected.

Generative research is about being inspired to create new concepts. Therefore, having the broadest range of participants will give your research team that broadest set of behaviors to observe and use for design inspiration.

Conducting the Research

Our studies are generally structured with an initial interview at the beginning of the study to get to know participants and build rapport. Often the middle part of a study will involve gathering data from participants through logs and voicemail diaries, and studies will conclude with interviews to clarify the details of data collected throughout the study and to ask any final questions. Most of our studies run for about two to four weeks so that we can observe a variety of behaviors that occur and begin to understand interactions that are frequent, while still having the opportunity to learn about more rarely occurring but impactful events in our participants' lives.

We generally take two to three researchers into the field for any given field study. There are a few reasons for this. First, we would like as much of the team as possible to get to know the participants in the study and we would like the participants to get to know as many of us as possible. Grounding researchers in at least some details of each participant creates a familiarity that is useful during analysis. It also helps us build rapport with the participant and introduces them to anyone who might call during the study to remind them to file a voicemail probe or schedule a final interview.

Second, having two researchers allows for more of a dialogue with the participant and allows for someone who is not focused on coverage of the primary questions to focus on additional follow-up questions based on the participant's responses. This can take some pressure off of the primary interviewer and almost always leads to more complete data.

Finally, we generally audio- and videotape all of our interviews so that we do not have to frantically scribble notes during the interview itself. The second or third researcher can handle the technology and make sure everything is captured. This is especially helpful during interviews that contain home tours as the visual record is critical to the analysis of this portion of the research.

We also strongly believe in conducting interviews in person. While this may involve extra time and travel expense, meeting with someone to establish a relationship and seeing his environments, media, and devices provide countless benefits in interpreting the data that is collected in the context of his life. We also often ask users to bring out specific devices or media that they mention in an interview, and this would be more difficult over a distance. Some promising work has been conducted using Skype video for conducting research remotely, but this requires users with a bit more tech savvy and still does not afford the ability to look around a participant's environment for additional cues.

We generally start a new research project with one to two pilot participants who are usually not part of the reported data. This allows us to test our research protocol, add additional questions, check the length of interviews so that we can report approximate time commitments to participants and recruiters, and ensure that our research questions are being answered by the data that is collected. After a pilot study, we will often rework parts of the study in preparing for the actual set of participants. Small changes can still be made once a study is in the field, and we generally debrief as a team after each interview in order to update questions and discuss follow-up questions that we should add to the protocol based on the initial interviews.

Affinity Analysis

Often, a generative study can create thousands of individual pieces of data. That data could be a sentence or two that a participant says that pertains to a particular research question, a photo, an observation made by the research team, or any other information that the team has collected that pertains to the questions that are to be answered. We prefer to conduct our analysis directly from the raw data instead of using researcher summaries or insights as a base. This gives us the ability to trace specific higher level findings back to specific users and statements and to be more confident that the findings result from actual actions that occurred and not from biased researcher notes that might be more selective in nature or more likely to interpret specific actions and quotes from users in particular ways. With all of the raw data in the analysis, the traceability back to the individual participant is always present should anyone question (or just want to better understand) a higher-level finding.

The process of getting from collected data, often in the form of audio or video files of interviews and voicemails, to data items for analysis involves some effort. We first listen to each recording and transcribe any bit that pertains to one of our research questions. Each relevant statement from a participant gets its own item for analysis, often just a sentence or two long. On the item, we note the participant that the data came from and in what context (Initial Interview, Voicemail from day 2, etc.). We also often note the timecode in the file where that statement occurred in case we want to pull out video or audio for later presentations. Each item is also numbered for later identification and printed onto a post-it note for analysis as shown in figure 3.3.

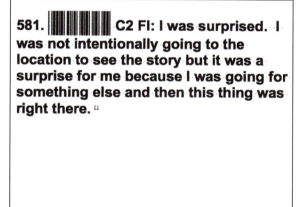

Figure 3.3

Example note from an affinity analysis on the Serendipitous Family Stories system. The note has a unique number, a bar code representing that number for easy scanning, the participant ID (C2), and the part of the study from which the data was collected (FI = Final Interview). This is followed by the direct quote from the participant. Hundreds (or thousands) of notes like this are then combined into an affinity analysis as shown in the following figure.

Once we have all of the raw data identified, we perform a grounded-theory affinity. The affinity is a bottom-up approach for iteratively making sense of raw data and developing theories of interaction based in the raw data. This analysis comes from the combination of methods from two different backgrounds. Jiro Kawakita, a Japanese anthropologist, created the KJ method of analyzing data (Scupin 1997). Seeing that standard ways of analyzing data in terms of existing theory and hypotheses were not working, he saw this method as a way to inductively build up meaning by finding interrelationships between items of data.

This affinity analysis method was popularized in the Human Computer Interaction community by Beyer and Holtzblatt (1998) in their paper and later book on contextual design. They adapted the KJ method and shortened the process by performing analysis on researcher observations instead of direct data from participants. We prefer to use the raw data for analysis so as to better control for researcher bias and to base findings in the actual quotes and data from our participants.

In an affinity analysis, the first step is to group data based on similarities that are observed. These first level groupings seek to describe the data and patterns that emerge across users who are saying or doing similar things. These similarities might be anything: two users trying to perform a similar action, examples of how place and other contexts shape an interaction, similar concerns or praises that participants raise, etc. These groups can represent any relationship within the data that the research team thinks is interesting and often evolve many times as the larger data set is analyzed. Often during the analysis a group that started small will grow as more supporting data is found. These groups are then broken down as nuances between the data items are more apparent.

Figure 3.4
Photos from an affinity analysis showing the hierarchy that is developed as the data is analyzed.

In the end, we try to get the first level groupings down to piles of three to six notes from a variety of participants. At this point, we create an explanatory label on a new, differently colored note. This label serves to state the main message of the group, often in the first person. Examples of labels from a study on communication over a distance (Bentley, Harboe, and Kaushik 2010) include: "Gifts from my kids around the house remind me of them"; "Family hand-me-downs around my home remind me of my parents"; and "When I wear jewelry from my parents, I feel that they are with me." Each of these labels had between three and six items supporting them, generally from multiple study participants. When a variety of diverse participants start to say the same things or do similar actions, it becomes more likely that we have found something more generally applicable to a larger population instead of just the idiosyncratic behavior of one or two participants. In some studies we have seen seventy-year-olds and twenty-year-olds using communications technology in similar ways which helps us know that we are on to something much more generally useful.

The second level of grouping in the affinity hierarchy seeks to explain the data. Groups from the first level are combined together into sets that tell a larger story about the behavior that was observed. In the case of the three groups mentioned above, this could be a statement such as "Physical objects can create a sense of togetherness over distance." These larger groups help the research team to look beyond the individual notes and on to patterns of use that occur at a broader level. These groups can be seen as sub-findings of the larger research: findings that have some amount of support behind them from several clusters of lower-level notes. In a study with 1,000 data items, we can typically have twenty-five to thirty of these groups.

Finally, a third level is created that serves to tell a larger story of use. Usually a given study will have a small number of these high-level themes that can serve as categories to summarize the major findings in the study and as themes to work off of for inventing new concepts and validating existing ones based on the data that has been collected. Some examples of these high-level themes in our intergenerational communication study include "Sharing stories about daily life creates togetherness over a distance" and "The concept of the family is an important part of family communication." These themes helped us see the larger goals of our participants in maintaining their relationships across a distance and what they did to keep these connections strong when they could not see each other in person as often as they would like.

Once all three levels of the affinity are complete, we hold brainstorming sessions to invent new concepts that are based in the data. We place short descriptions of each idea on brightly colored post-it notes right next to the raw data or insight that inspired them. From a study with 1,000 data items, we often create in excess of 100 design ideas. These ideas are directly tied to the themes and participant quotes in the affinity diagram. Examples such as the Metadata Knob came from observing a user trying to find music that was a bit faster than what he was playing and another user who was searching for something a bit newer than the currently playing song. The idea for Serendipitous Family Stories/StoryPlace.me came from a similar observation. One of our participants was walking past a theater and was talking with her mom about the importance of that place in their family history. We saw the opportunity to create these types of interactions serendipitously with location-based videos and notifications when users approached locations that contained videos that were shared with them. The ultimate concept used each of the high-level themes from the analysis to create a system that reinforced the memories and roles of the family, created communication that fit into everyday life, and created a feeling of being together with family members by placing video into real-world settings. While there are certainly other ways to arrive at concept ideas like these, we have found that the affinity analysis and using actual experiences to ground new concepts gives us the additional confidence that a new idea is useful for a broader population than ourselves. Also, the sum total of the data in the affinity and the resulting design guidelines help shape our new concepts far beyond the single note or group that inspired them. This allows the resulting solutions to more accurately fit into people's everyday lives and interactions with the world.

While sometimes we have one or two ideas that everyone agrees are absolute winners, often a prioritization process is necessary in order to identify the ideas that are the most valuable to take forward. These prioritizations can contain columns for applicability to current businesses and products, novelty/patentability, difficulty in building the concept, potential market size, and other factors. From this process, we generally have two to three high-quality ideas that we wish to take forward into functional prototypes. We have also found that the design guidelines generated can be quite helpful to other product teams developing concepts in related spaces (Metcalf 2011).

The next step for the highest-rated design ideas is often a quick prototype to validate the design hypotheses as well as some technical aspects of how they could be built. That will be the focus of the next chapter, but first we will focus on some ways to shorten the process of generative research when time does not allow for a more formal study.

Discount Methods

Often, research or product teams do not have time to complete a full, rigorous study but still would like to base their designs in user observations. In our class at MIT, students have a week to run a quick study and only an hour or two in class to analyze the data that they collect. This does not allow for time to create a formal interview protocol, gather and transcribe hours of interviews, and create 1,000+ note affinities. But we have found ways to make the short time that they have worthwhile so that it improves their overall designs and grounds their systems in real-world behaviors and needs.

The design firm IDEO (Kelley, Littman, and Peters 2001) popularized the concept of a "deep dive," a quick way to put designers in contexts of real use, both by observing people and figuring out firsthand where existing products and environments failed their users. Their video from ABC's Nightline (Deep Dive 1999) in which they design a shopping cart is required viewing in many design schools and HCI classes around the world. Getting out in the world with a quick observation is a great way to ensure that new concepts are grounded in everyday realities. We encourage our students to take a bit more disciplined approach that lies somewhere between the types of observations shown in this video and the full, more ethnographically inspired, rigorous process mentioned above. First, we start with a list of research questions that get to the core of what the team wants to discover. From there, we ask them to spend an hour with people from their target audience watching what they do and then interviewing them to better understand their behavior. They can also ask participants to log behaviors or take photos in a log for five days. The data collected (often in the form of observation and interview notes) is then written down on post-it notes. From this hour of observations, students frequently have between fifty and one hundred notes. We then perform a mini-affinity in class where they group the notes to find patterns and explanations of use.

While a one-hour observation is not going to be as exhaustive as a full study with a dozen or more participants for several weeks, it tends to be a good start and gets students thinking about their target domains in new ways through the eyes of the people that they observed. Often, this leads to follow-up observations or the desire to observe more people in different contexts of use. In almost all cases, it helps students to see how some of their initial assumptions might not have been correct and that the range of behaviors in their domain of interest is different from what they imagined. This helps them to create stronger concepts that will work for a broader range of people, which is ultimately the goal of generative research. Some of our students in the past few years have explored a variety of routine domestic activities. In each case, by observing and talking with just a few people over an hour, teams were able to identify a range of behaviors much broader than what they had originally expected. In each case, ideas from these observations would become key differentiating components of their final solutions. Often, students find different types of user segments that they had not expected or interesting ways in which people are already trying to accomplish a task. These insights dramatically improve the quality of their final systems and create solutions that are likely to work with a wider population than what they would have created without this generative data. Also, as in the shopping example discussed earlier, students often discover a key part of an activity that makes it fun and can turn that insight into a focus of their ultimate design.

Other methods, such as those that IDEO has outlined on their Method Cards (2002), can help to quickly explore current practices in a variety of scenarios and lead to quick observations or interviews that can ground a design in real-world behavior. Most of these methods can be completed in a few hours and can give design and research teams quick new perspectives on their domain of interest.

However, researchers must take care to understand that the data they collect in any of these rapid methods is just a small sample of the data that it is possible to collect. If only a few people are observed, it is easy for research and design teams to overly constrain their resulting design for the types of use that were observed and thus leave out many potentially common use cases. This can also happen with the use of personas or user-types derived from small amounts of data. Designers often focus on designing for those particular personas and forget that other types of users could exist that are hybrids of the personas or that represent entirely different types of use. These discount methods should be carefully used to broaden design thinking and open up new possibilities instead of narrowing resulting designs to conform to the small amount of data that is collected.

4 Rapid Iterative Prototyping

Just because an idea is inspired from user data does not necessarily mean that it will become a successful product or fit into the daily routines of those who use it. The only way to know how a new application or service will be used in everyday situations is to build a critical part of it and see how it is adopted in use over time. While sketching an idea on paper or having people "experience" a storyboard or paper prototype in a lab setting is often a first step that many groups take, these types of prototyping are not sufficient for understanding how a new service will be adopted in daily use and can sometimes even be harmful by soliciting opinions and initial thoughts about a concept that might differ from those gained through actual, sustained use.

People are inherently bad at picturing how their lives would be different with a new application or device. In the late 1990s, studies were conducted to understand why some people were purchasing Internet access and some people were not (Shirky 2008). Those not online were asked what they thought the Internet would be good for and in general they replied that it would be good for looking up information. But once people got access to the Internet, they realized its tremendous social potential through applications such as email or instant messaging and those uses dominated their time online. As demonstrated when more people came online, the initial non-users would end up primarily using these communication features that they could not fully understand the benefits of until the features became a part of their lives. Similar to this, a focus group or a look at storyboards will not give people the insight they need to understand how they would use a new product. Their responses will be grounded in their existing knowledge and current behaviors, not in actually using the system and experimenting with it to adapt it to their own needs. Even using a new concept in a lab setting will not allow people to fully understand how it could be used in their everyday contexts. Using a system for a few minutes in a lab setting is very different from having it available at any time during which multiple new uses for the system can be discovered as a user goes about his or her day.

Because mobile systems are used in a variety of contexts and for reasons not fully anticipated by system designers or novice users, it is important to have a working version of a system for a user to interact with in the context of their lives for a period of time. Having a system always available will allow users to reflect on ways in which the system can fit into their lives and the system creators can observe a wide variety of interactions and tasks that their system supports. This understanding can lead to a stronger focus on key user values of the system as well as help make decisions as to whether to continue on with a given concept or move on to another idea.

Only trial and error and experimentation in daily life will show how new concepts fit (or do not fit) within people's lives. We often build out key pieces of our new experiences within weeks of getting an idea and quickly get them into hands of users as a means of risk mitigation before too much effort is spent on a given idea. The earlier in the process an experience is realized and tested, the less time is wasted pursuing ideas that will likely not succeed.

In most new concepts, there is usually a core bit of the idea that contains the main new experience and is what needs to be tested in order to determine if the experience is worth pursuing further. Questions often arise early on about how a part of the concept will fit into someone's life, how particular data might be used, or if a given algorithm reliably provides the data that it should to help people in a given situation. In addition, questions around performance and usage models are often critical to establishing a business model or appropriately scaling a back-end solution. Examples of these core experience ideas from research at Motorola include: moving/not moving data in a phone book, videos tagged with a place and shared with friends who discover them at that place, and viewing multi-angle video instant replays of sports events on a mobile phone after each key play in the game. Examples from our class at MIT include systems that let users connect with the right group of friends to do a particular activity as well as collaborative to-do lists.

The prototypes that we develop to answer these research questions often start out as a small part of the overall system and lack the full functionality required for a commercial product. We use them to test out the key technical and behavioral research questions that are core to determining the viability of a concept instead of building a more full-fledged alpha version of the complete idea. They are often messy with lots of settings hard coded and without user-configurable aspects that are not critical to exploring the new concept. Because they are simple and need to support only dozens of users for our initial trials, it often only takes a few days or weeks to get them built and into the hands of participants, thus providing not only feedback that is more reliable than other early-stage methods such as storyboarding and focus groups, but also a first tangible version of the concept that others can touch and use in their lives. This is extremely valuable in explaining the concept not just to potential users, but also to investors, marketing professionals, or sales teams who might need to be bought into the concept before a full commercial system can be built.

We have three main principles that guide us when producing an early-stage concept to test in the field (Bentley and Metcalf 2009a). These principles help us quickly create a functional application that participants can use over the course of several weeks in their daily lives so that we can learn how it is adopted into daily practices. These principles are: *build only what you need*; *build the experience, not the technology*; and *build it sturdy (enough)*. By following these guidelines, we ensure that we spend the least amount of time possible on a given project before validating the key components of the experience. If the idea does not turn out to be a good one, we have wasted little time on it.

Within this process, it is also important to ensure that the concepts that are created are usable and understandable by novice users. We often perform rapid versions of some of the methods described in chapter 6 at this stage to quickly test parts of our initial prototype designs. Techniques such as heuristic evaluation or paper prototyping/ rapid usability testing can help make sure that what you are about to field is usable. However, since these initial concepts are often quite simple and contain just a small set of screens, these methods are often not needed at this stage. Examples in the remainder of this chapter will illustrate the relative simplicity of design needed for an early-stage rapid concept prototype.

Build Only What You Need

Because we want to quickly evaluate a new experience, our prototypes tend to be made rapidly, often in a matter of a week or two, and are built to test a specific aspect of a concept with users. We prototype the key use cases and leave the less-relevant functionality out. What is implemented in the end still needs to be a usable system in daily life, but it does not need to include all of the added functionality of a complete application that can be publicly released. Those details can be added at a later time if it is decided that the new concept is worth pursuing.

In deciding what to build, it is necessary to review the specific research questions that you have about a particular concept. These often fall into two buckets: questions about the technology and questions about the use of a new application or service. Often, a single prototype can help to answer both types of questions.

In early 2006, we had an idea for a new type of location-based presence on mobile devices (Bentley and Metcalf 2007). After Frank and Crysta Metcalf (2008) had completed the location sharing study and had read several papers about unique cell tower identifiers and their density in both urban and rural locations (see Hightower et al. 2005; Smith et al. 2005; Sohn et al 2006; and Chen et al. 2006 for details), we had the idea that switching between cell tower identifiers could help to determine if a user was at a place or moving between places. This type of abstract information seemed perfect for presence information since the data by itself (moving or not moving for a particular number of minutes) does not give away much information, but requires additional social knowledge that close friends and family likely posses in order to be translated into something meaningful and used for awareness/planning/etc.

For this concept, we had several research questions. First, we wanted to know if some basic algorithms that we created would work for determining moving and not moving states. We wanted to know how our derived measures of moving and not moving corresponded to a user's mental model of these concepts. Finally, and most importantly, we wanted to know if this data could actually help someone in their communication and awareness of close friends and family members.

The system that we built was extremely simple. We created a basic modified phone book application that read the names and contact numbers that were to be displayed from a file on the phone. It ran a background process to monitor cell tower identifier changes and, when a user transitioned between moving and not moving states, a text message was generated and sent to a hidden port on the phones of close friends and family participating in the study. This message was never displayed to the user, but intercepted by our phone book application on their devices to display the appropriate number of moving or not moving minutes in the interface. All in all, it took about two weeks for one person to build and refine the application and motion algorithm while attending to other activities in the office.

More important than what the application did was what it did not do. There was no way to add or modify a contact. There were no settings for customized ringtones or photos or any of the other bells and whistles of a modern mobile phone book in 2006. Likewise, because all data was sent via text message from peer to peer, we did not need to create a server or manage finicky persistent connections.

This rapid prototype allowed us to quickly see the power of the motion presence concept and the many uses to which our participants were able to apply the data. They used it for meeting up with friends at the same place and time, for knowing if it was okay to call someone because they were driving or already in the office, for knowing that a loved one was safely at work or home, and in many other ways. Just a very simple and crude-looking prototype can be all it takes to see how a new concept

 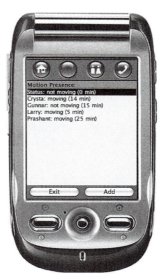

Figure 4.1
Very rapid prototypes of the Music and Motion Presence applications. These systems were created in days or weeks and allowed us to quickly learn about how the concepts could be used in daily life. (Images used with permission of Motorola Mobility)

fits into people's daily lives and interactions with each other and the world. A deeper discussion of evaluation methods is the focus of chapter 8.

Our Music Presence application is another example of this philosophy (Bentley et al. 2007). Created by Drew Harry while a summer intern at Motorola Labs, the system was built to answer research questions about what people would do if they had a real-time feed on their phones of the artists and titles of music that their friends were currently listening to. To answer these questions, Harry made a simple prototype that polled the Audioscrobbler website API (2011) for music updates from participants and text messaged these updates to their friends' phones. The system was up and running in days and we were able to field it and learn quite a bit about how music information could be used to infer when someone was home, bored on a Friday night, or in a particular mood. Receiving this information in an SMS was certainly not the best user experience, but we could quickly learn if there was anything useful in the idea itself. The concept showed significant promise and so we continued on with further prototypes, eventually leading to many of the concepts in the Motorola Connected Media Player (FLIPOUT 2011) and Last.FM (2011) integration into the phone book on MotoBLUR-enabled Android mobile phones (Bentley et al. 2010). We could not have learned this information through paper prototyping or lab studies; it was critical to build a functional experience and test it in the world with real participants.

These prototypes will likely look quite crude to anyone used to creating consumer applications. By design, we do not spend time on professional-looking graphics and nice layouts at this stage. As Rettig (1994) pointed out when conducting usability studies, participants will focus on the small details of particular graphical elements or not want to be too critical of an overall concept if they see a "finished" design. With these very basic prototypes that we have fielded, it is very clear to the participants that we are testing a new concept, something that is still very much in flux, and that their feedback will help us turn it into something better.

Build the Experience, Not the Technology

Because these early prototypes are often focused on answering research questions about how a new system will fit into the lives of our participants, the prototypes that we build are often not engineered in the way that a commercial offering would be. Because we utilize small-scale studies with limited numbers of participants, we have the ability to quickly put together a system that is good enough to work in that environment.

A prime example of this came with the creation of the mobile sports video solution TuVista at Motorola Labs (Bentley and Groble 2009). In the fall of 2008, we wanted to create a system that would allow fans in stadiums to receive video instant replays and stats from the game that they were attending. We hypothesized that this would add to the enjoyment of the in-game experience and that fans would replay key events in the game both during the event and at natural breaks, such as half time and post-game. But when we had these ideas, we did not have a system.

We quickly created a system using mainly off-the-shelf components. We set up a computer with the commercially available video editing software Final Cut Pro to make video clips during the event and created a simple website to which the producer of the event could upload clips and enter stats. A minimal Windows Mobile application received updates over a multicast channel when new content was available and let users scroll through all of the content from that event.

None of these technology decisions would end up being the way that we would make our commercial system, but they all served perfectly well to understand how people in a stadium would react when provided with instant replay clips and stats as the game progressed. We were able to quickly set up the system in Estadio Azteca in Mexico City and ran a trial with sixty fans at the South American Cup Semi-Final

Figure 4.2
The first prototype of the TuVista system used off-the-shelf components such as Final Cut Pro, a simple website, and a basic mobile application so that we could quickly understand the use of video replays in the context of a sporting event. (Images used with permission of Motorola Mobility)

game in 2009. We observed a group of users celebrating after the game, playing the video of the winning goal on their device, which was held high in the air, while they danced and sang the team victory song. We also collected usage data and saw peaks of interaction in down times in the action on the field. The insights learned from this study on the use of media in groups, the desire for content to appear almost immediately after a play, and the use of breaks in the action to relive key exciting moments in the game would help us design the next version while we simultaneously attacked the technical issues of rapid multi-angle video editing, multiple device support, and supporting multiple networks and devices behind a Network Address Translation (NAT) not reachable by multicast.

The final solution that we took to market included a very different tool with which professional content producers can create videos clips, bundle them, and publish them to devices as shown in figure 4.3. This was created with the results from our initial exploration that showed us where we would need to focus to make a differentiated solution and where bottlenecks existed in current systems. Our end solution was the fastest way available to publish multi-angle clips from live events for years after its initial launch.

In the Motion Presence application mentioned earlier, we simply sent SMS messages to a particular port on the phones of friends and family any time the motion

Figure 4.3
The commercial editing tool for the TuVista product was created based on insights gained from initial rapid prototyping and field evaluation work. Our focus on quickly editing multiple clips and applying metadata to sets of content made this interface the fastest service available for content producers publishing clips from live events. (Image used with permission of Motorola Mobility)

status of a friend changed. This could be implemented quickly and because we were using small groups, it worked. However, as each user changed motion status about fifteen times per day, this resulted in 3,360 text messages being generated over three weeks in just one group of four participants. All of our participants had unlimited text messaging plans, but such a solution would not be suitable commercially and could risk being shut down by network operators as their networks would be flooded with thousands of additional messages. A commercial solution to this application might tie into more complex Internet Protocol Multimedia Subsystem/Session Initiation Protocol (IMS/SIP) presence systems and would have taken much longer to implement.

In the Serendipitous Family Stories (Bentley, Basapur, and Chowdhury 2011) prototype (shown in figure 4.4), we wanted to create a system for older adults to record

Figure 4.4
The web and mobile interfaces for the Serendipitous Family Stories system. We implemented only what we needed for the field trial and did not include any account management or friend finding functionality that would be required in a real public release. (Image used with permission of Motorola Mobility)

video stories about particular places, which they could save at a point on a map and share with their relatives. Their relatives would have a mobile phone that would monitor their location and when, as a part of their day, they came across a story location, the phone would vibrate with a unique pattern and allow them to watch the video in its physical context. Sujoy Kumar Chowdhury, an intern at Motorola Labs, and Frank implemented only what was needed to field the system. We used free versions of Adobe's Flash Media Server for content creation as we knew that we would never have more than ten simultaneous streams. We also did not include the authentication and Application Programming Interface (API) security mechanisms that would be needed for commercial systems. In the end, a simple Representational State Transfer (REST)-based API that provided direct access to database resources provided an uncomplicated means for mobile applications to access story lists and details for stories that were shared. A simple Hypertext Preprocessor (PHP)-based web interface provided the basic functionality that was necessary for the user trial but did not offer any account management features or more complex sharing behaviors. Users in the study were only sharing content with one person, so this functionality was simply not needed. Final database schemas and system architecture would vary greatly in a commercial solution.

Another common way to reduce the amount of work that must go into an early prototype is to create a mashup or incorporate existing web services into the design to achieve the desired experience. For Music Presence, we used Audioscrobbler to receive music updates. While ultimately we saw the value in collecting this information ourselves, using Audioscrobbler saved us many weeks of implementation and included off-the-shelf plug-ins for the most popular music playing applications of the time (e.g., iTunes, WinAmp, etc.) with no additional work. All that we had to write was a simple script that polled their APIs instead of a full music-tracking solution. We combined this with the SMS system on the phone to deliver updates so that we did not have to write any client code in order to display music updates from friends. A simple server-side script could handle all of the API-polling and delivery in order for us to see how people would interact with this data and what it could be useful for. As in the Motion Presence system, a very large number of SMS messages were generated. But since we were just conducting a small study with users who had unlimited SMS plans, this was not an issue in getting us the data we needed.

We believe that testing a quickly built version of an overall concept is the most important first step in understanding which concepts to pursue for further research and commercialization. If the concept fails, there is often no reason to spend the time to develop the technology further to produce a more scalable and secure solution. And for concepts that prove successful, a quick throwaway prototype can often identify the technical bottlenecks that might need to be improved on the way to a commercial version. In his still-relevant book on software development, *The Mythical Man-Month*, Fred Brooks (1995) discusses the concept of building one system to throw away while taking the insights from this system forward to a second system. We wholeheartedly agree. Not only can a first system help researchers to learn about the usefulness of a given concept, but developers can also learn quite a bit about areas that will need further work. Our TuVista system showed us how much of a bottleneck linear video editing really was and caused us to invent a totally new solution for editing multiple simultaneous video streams. The Serendipitous Family Stories field trial moved our attention to the larger issues of authentication, sharing, and data privacy concerns.

Build It Sturdy (Enough)

Even though we focus on building quickly, mobile concepts that are going to be used in the field need to be built sturdy enough to survive everyday use for multiple weeks by people not familiar with the technology or interaction. If an application frequently crashes or does not have up-to-date data, users will often stop using it or otherwise change their behavior. Building minimal systems, as described above, makes it much easier to reach the point where additional work would not make the application significantly more robust. Likewise, we find that an internal pilot is almost always necessary to iron out any kinks that arise when taking a concept out of the lab. It is much better for a research team to catch these issues than to learn about them during a deployment and need to throw away data or repeat a field study after a bug has been fixed or complex interaction flows improved.

This generally means avoiding the use of new or untested technology at this stage of the research unless it is critical to the research questions that need to be answered. In times when we relied on working with advanced technology components or fully developing an interface, we often found the key concepts of the system to be lost in the complexity or that the new technology did not perform to our needs for the given experience. When this happened, we went over our anticipated deadlines, and in one case, had to stop a study entirely because of the state of the underlying technology. We would much rather test the concept using alternative implementations and take the time to more completely develop the appropriate commercially viable technology once the concept is more fully validated.

While it is often not necessary to conduct a full usability test at this stage, having a few people try out the system is rarely harmful. Often a simple change in the text of a label or movement of a few User Interface (UI) elements can greatly increase the usability of a mobile application. Since the first rapid prototype is usually fairly simple in terms of a feature set, a quick heuristic evaluation of the interface and a run-through by a few non-engineers or designers is often sufficient. Some useful heuristics are available from Nielsen (2005) and Wodtke's (2002) eight principles. These heuristics ensure that interfaces provide appropriate context for users as they seek to understand and navigate an interface and also that appropriate error cases are handled gracefully. Generally, it is best to think about these heuristics while designing the experience, but it is always beneficial to have a trusted colleague who is familiar with them perform a quick review of your system before you put it in front of users.

There are some common components of mobile applications that generally lead to issues in the field, and addressing these early on in prototypes can help make the experience of running it in the field much easier. The most common is gracefully handling intermittent data connections. As users go about their lives, they move among a variety of networks. At some times they may be on Wi-Fi, at others on a 2G or 4G (Second or Fourth Generation) network, and yet others in no coverage at all. It is common when moving between networks for phone platforms to drop all existing data connections. Mobile applications should be aggressive about retrying failed server interactions whenever a data connection is present.

Because users likely will spend at least some part of their day outside of network coverage (e.g., in an elevator or subway), it is important to consider offline usage for many applications. Appropriately caching data so that applications are usable in these cases is important. The data might not be the freshest or most contextually relevant, but it often will still be useful and better than displaying a blank list. If users travel internationally, they often will have data turned off to avoid excessive roaming costs. In these cases, whatever is cached is likely all they will be able to see until they encounter some Wi-Fi. It is important to think about these use cases before giving participants an application to use for a few weeks or months. In Motion Presence (Bentley and Metcalf 2007), all data was cached on the device and sent through text messages which would be automatically resent by the network if a recipient was out of coverage. In Serendipitous Family Stories (Bentley, Basapur, and Chowdhury 2011), all data was cached in a local database so that stories could be discovered and the story list would be visible even when no data access was available. This also made the system faster as these screens were always available without further data connections being necessary at the time of interaction.

Another common area that needs hardening before taking a concept into the field is text entry. Following best practices to properly encode non-ASCII (American Standard Code for Information Interchange) characters is quite important when testing a concept in the world where users expect to be able to enter any text they want into any free text field. Properly encoding characters for Hypertext Transfer Protocol (HTTP) calls and Structured Query Language (SQL) database storage is crucial as they are the two most common places where errors occur with nonstandard characters in rapid implementations.

If an application is going to be fielded on multiple types of hardware or networks, often some reformatting of data is required. For example, if an application is using platform APIs to receive a phone number, some carriers' Subscriber Identity Module (SIM) cards report this with the country code while others do not. In the Serendipitous Family Stories prototype, we had to check for this and add the leading "1" when omitted (all of our users had U.S.-based phone numbers) in order for content to be appropriately shared between users. Likewise, different versions of Android technology can use different location providers (SkyHook, Google, etc.) that could provide different location resolutions in particular areas. Some of the differences in location technologies will be discussed in the next chapter.

Other Pitfalls

While we have seen this method of rapid prototyping and evaluation work countless times in research at Motorola and with our class at MIT, there are definite pitfalls to be considered when applying this approach.

Since the studies that are conducted with these very limited prototypes are by necessity small in terms of the number of participants, there is always the problem of generalizability of the data that is obtained. As with much rapid ethnographic-style research, this can be mitigated somewhat by selecting participants that are quite different from each other and observing similar behaviors across these participants. Often, we explicitly try to recruit participants across a range of income, education, race, and gender. If we are testing a particular system that augments a behavior that they already exhibit, we look for a wide range of existing behaviors present during that activity (e.g., behaviors such as hours spent watching television or number of phone calls made per week). This helps to get a more representative sample.

However, no matter how representative the sample, the data that comes out of this type of study will not inform you as to the expected frequencies of any type of observed behavior, just that certain types of behavior are possible. Other methods with larger sample sizes are necessary to get more market-specific data on segmentation and feature use. The data captured can only help to build initial usage models, to be refined with more detailed marketing data as the concept progresses toward a product.

Another problem that often arises is the lack of critical mass in small-scale deployments. If only ten people in the world are using your system, it is hard to see how social interactions will play out or how communities will form in social systems. For systems that rely heavily on mass social participants, a larger alpha release might be necessary. However, many systems for mobile devices are meant for use in smaller social networks of friends and family. For these types of applications, we have been successful in recruiting groups of friends and family to participate in our studies, thus achieving some critical mass among a group of people that is most likely to use that particular system together. We have found that groups of four or five close friends or family members tend to be a good size for these types of studies. That number provides enough people that social interactions reach multiple people for whom the participant cares strongly and is high enough that many interactions will have relevance to at least one of the other participants.

Also, the quick and dirty prototype-quality implementation of the system can often alter the way the system works in practice. Interactions may take longer than they would in a commercial-quality system. All of the data may not be present and users may not be able to perform key tasks that have not been implemented at this stage. It is important in the analysis of the data that is collected to understand what particular usage behaviors were a result of the limitation of the system versus the concept itself. While noting usability issues, it is more important to focus on use of the system, the particular tasks that users accomplish with the system, and what their motivations are. This focus helps to discover what the concept is good for without overly focusing on the implementation-specific technical or interaction issues.

These guidelines have allowed us to be quite rapid in discovering key experiences that are meaningful in the lives of our users. Often, when we finish an exploratory study, we have hundreds of design ideas from which to choose. Taking a few of the most promising and quickly prototyping them and testing them in the lives of users helps us to identify which are the most promising to pursue for further research or for commercialization.

Discount Methods

While building a quick prototype in a week or two and testing it in the field over several weeks is generally a quick process, there are ways to make this part of the process go even faster. In many cases, just to test an experience, following the adage "If you can't make it, fake it" is appropriate and can still let you explore a concept in real-life use. These discount methods often involve a technique known as Wizard of Oz. Like the famous Wizard in the book by L. Frank Baum (1900) who stood behind the curtain pulling levers to create the illusion of the "Great and Powerful Oz," Wizard of Oz studies involve a member of the research or development team acting out a part of the system behind the scenes. This can save valuable time when implementing an algorithm that may not actually work in daily life. However, a Wizard of Oz system often requires a great deal of work and attention on the part of the research team while a study is active in the field, especially if it is one that occurs over a period of time in daily life where interactions can happen at any hour of the day.

Trevor Darrell and his student Tom Yeh at the MIT Computer Science and Artificial Intelligence Lab created a mobile tour guide system for the MIT Campus (Yeh, Lee, and Darrell 2008). They loaned out phones to visitors and invited them to take pictures of places they would like to learn more about. Since they could take a picture of anything, and since the application needed to work indoors where location technologies could not provide detailed information about exactly what a camera is pointed at, they decided to provide the information themselves. As soon as a user took a picture, researchers would receive the photo and could send back links or text relating to the object that was photographed. To the end user it appeared as if the system provided these results automatically. While this meant experiencing a bit longer of a delay than they would with a fully automated system, users could get detailed information and the value of the greater system could be evaluated without the need to implement fancy computer vision and localization algorithms. Those could be researched and developed should the concept itself prove useful.

Mashups of existing mobile and web services can also serve as a means for rapid prototyping. Henriette Cramer and colleagues at the MobileLife Centre in Stockholm were interested in ways in which people could explore music that was being played in various nearby venues. Instead of building a system from scratch, they used the Spotify and FourSquare APIs in order to create a service that links music being played to the places where people have checked in (Cramer, Rost, and Holmquist 2011b). With a simple mobile web application, they were able to leverage large amounts of data already being logged by existing systems and explore that relationship in a new way.

The most important part of building and testing a rapid prototype of a mobile system is to get out of the lab and into the world. Learning from interactions that take place in everyday environments brings insights that are not possible to gain when observing users perform predefined tasks in a lab environment or when users are constrained by the times and locations in which they can use a particular service. These rapidly built systems can not only test use in situ but can also help to determine the technical feasibility of various components of the planned system and help to plan for anticipated data use and storage.

5 Using Specific Mobile Technologies

While in many ways mobile devices are mobile computers, in key areas their additional functionality (or lack of functionality) strongly impacts the types of experiences they can enable. The environments in which they operate require careful use of location, network, and sensor APIs in order to conserve battery power while still providing a satisfying user experience. Often, those new to designing and implementing mobile experiences overlook these issues and the resulting mobile experience suffers. Mobile devices might be seen as always-on, always-available computers, but in reality they are still quite limited in terms of the persistence of their connections, their ability to obtain a precise location at all times, and their battery life. Following some of the guidelines in this chapter should lead to creating mobile experiences that are more pleasing, even in times of intermittent connectivity, and extend the life of a charge for your users.

Understanding these technology constraints is a critical step in the mobile design process. Often, designers who come from nontechnical backgrounds do not know the full implications of some of their design choices. When this occurs, the end-user experience frequently suffers. This chapter will provide a brief introduction to the specific opportunities and limitations that mobile designers and researchers must understand so that the experiences they create fit within the inflexible technology constraints of today's device.

The typical smart phone in 2010 had a battery capacity of around 1300 mAh. Every piece of hardware on the phone uses power when it is activated, and when designing a new mobile experience it is important to take these power drains into account, realizing that the user is likely running many other applications and services on their device at the same time. All of these power drains add up quickly on a modern multitasking smartphone. In an analysis of power consumption on the HTC Dream device, Amir (2010) found that when idle, the phone has a current drain of 83 mA. Therefore, using a 1300 mAh battery, it can last for 15.6 hours running the processor and radio. At

Table 5.1
The power used by various types of hardware on modern smartphones.

Device/Sensor	Power Drain
GPS	135 mA
Display	8 mA
CPU + Radio	83 mA
Standby (CPU off)	7.6 mA

Sources: Amir (2010) and Real Science.

"standby" with the processor off and radio on, it used only 7.6 mA, thus lasting for 171 hours. Processor use is one of many factors that can dramatically affect the battery life of a device. Ensuring that threads properly sleep so that the processor does not remain active is a critical part of designing any service that must run in the background.

The display can often be a factor as well. A typical image on the screen can draw 8 mA of power in normal operation according to a study by Real Science (Ion 2010). Other sensors and functionality that affect the life of the phone will be discussed in the remainder of this chapter. Choosing which functionality to use and at what interval to sample can be the key to creating an experience that fits into a daily charging cycle or not.

This chapter will address four of the main differentiating technologies in mobile devices and the implications of these technologies on the range of experiences that can be successfully created within their limitations. We will begin with the most commonly used and often misunderstood technology, location sensing. Location sensing on mobile devices is not instant and involves tradeoffs in terms of speed and precision as well as the operating environment (e.g., outdoors vs. indoors). We will then discuss data usage. Mobile networks might be seen as always-on and high speed, but even in densely covered urban areas, mobile devices often see poor data rates and long pauses in connections to the Internet. The chapter will conclude with a discussion of Bluetooth and environmental capture, such as ambient audio and accelerometer data. Understanding this information is critical to creating experiences that are possible within the technical constraints of today's modern mobile devices.

Location

There are many different location technologies available for mobile devices to use and today's devices take advantage of several methods to determine their location at any time. There are advantages and disadvantages of using each of these and it is important to select the method that is best for a given application. Some methods have very fast response times but higher uncertainty, while others are quite precise but are slow or require a great deal of power to use. Some methods do not work well indoors or use additional power when signal strength is weak. Privacy concerns about the types of data being sent over the network and to whom they are sent also can factor into choosing an appropriate location technology and sampling frequency. In the end, the choice can make the difference between an application that drains a user's battery in hours or something that comfortably works in the background throughout the day while preserving the privacy of location data.

The coarsest determination of location can be obtained from using cell tower identifiers. Every Global System for Mobile Communications (GSM) tower in the world has a unique identifier that is made up of four numbers: Mobile Country Code, Mobile Network Code, Location Area Code, and Cell Tower ID. These four numbers are available to programs on most modern cell phones and can be mapped to a physical location by consulting a location database such as Google (2010) or SkyHook (2010). These databases have been built up over years, largely by brute force, to create a mapping between unique cell tower identifiers and the latitude/longitude of these towers. The Location APIs on many devices today will take care of this mapping automatically if you ask for a location using a Cell ID or network provider.

Cell ID lookups consume the least amount of power and are the fastest of any mobile location technology. As a part of the process of staying connected to the cellular infrastructure, a phone always knows the identifier of the tower that it is currently connected to, and thus determining location is just a simple network call to map that tower ID onto its corresponding physical location. On some mobile devices, Cell ID mappings may be cached, further speeding up location resolution and saving the power that would be needed to make a network transaction to resolve the ID into real-world coordinates. However, Cell ID is also the least accurate of the location technologies. Cellular tower density tends to scale with population and building density, so in rural areas Cell ID localization might be accurate only to an area of a few miles in diameter while in a dense urban area it can often pinpoint location to within a block. In 2006, Intel Research (Chen et al.) investigated the accuracy of Cell ID location and found that in urban areas of Seattle they could determine accuracy to within 163 meters 95 percent of the time. In general, Cell ID–based location can achieve between 150 and 1000 m accuracy depending on cell density. The team at Yahoo! Research Berkeley (Ahern et al. 2006) termed this "Zip Code Level Accuracy," which is a good way to think about Cell ID–determined location since it scales with population density much like a zip code.

Because of the low power consumption and quick speed to determine a location, Cell ID is perfectly matched for two types of applications. The first includes applications that need to constantly monitor location throughout the day. This can include widgets that need to display location-sensitive information such as weather or traffic, or systems such as Serendipitous Family Stories (Bentley, Basapur, and Chowdhury 2011) that monitor location throughout the day in order to actively notify users when they approach the location of a family story that has been shared with them. The second type of application is one that conducts location-based searches such as those for nearby theaters or restaurants. These applications do not need a precise location, but often need to respond quickly to display the relevant nearby results, and Cell ID localization's low latency and low power consumption work quite well in these cases.

Applications that use Cell ID can also preserve the privacy of location information by keeping it on the handset. If the mappings between frequently visited Cell IDs and their actual locations are kept cached on the mobile device, location lookups need not incur a network transaction, thus keeping the location data locally on the phone. This can be beneficial, as many users still do not feel comfortable with their devices continually reporting their location to network services. Mappings can be made to physical places via latitude and longitude coordinates or Cell IDs can be mapped to places manually. In Sohn et al.'s (2005) Place Its system, locations were identified by users (such as "home," "work," or "school") and the phone would find the Cell IDs visible when the user reported being at that place for the first time. On subsequent visits, the phone could determine that a user was at that given place again by simply comparing the current Cell ID to the stored places, not needing any network transactions. The Place Its system then allowed for reminders to be embedded in these locations such that users would be reminded of things they had to do upon arriving at a particular place, such as taking out the garbage can when you arrive home or sending an important email when arriving at the office. As of 2011, this functionality is now a part of Apple's iOS operating system (Friedman 2011).

However, since an initial network request is necessary in order to determine location on most devices, the Cell ID method can use significantly more power when signal strength is weak and more battery power must be used to transmit periodic location requests. Caching locations, in addition to providing privacy from location requests being sent over the network, can also help with this power usage problem as resolving Cell IDs to locations can then be performed locally for cached locations that are frequently or recently visited. This can dramatically increase battery life, especially when signal strength is low. It can also allow for locations to be determined in areas

Table 5.2

Properties of various location sensing technologies common on today's mobile phones.

Technology	Accuracy	Time to First Fix (sec)	Power Consumption (mA)
GPS	10 m	65	135
Wi-Fi Positioning	40 m	5	Variable depending on sampling frequency, network conditions.
Network Positioning	300 m	<1	Variable depending on sampling frequency, network conditions.

Sources: Intel, Microsoft, USCD.

that might have a signal but no data connectivity such as some subways, crowded sporting events, or concerts.

The next most accurate location technology is Wi-Fi Positioning (sometimes abbreviated WPS). In concept, Wi-FI works quite similarly to the way that Cell ID location works. Mobile devices can scan the available list of Wi-Fi Service Set Identifiers (SSIDs) and use an online database to map the list of visible SSIDs to a set of coordinates in the world. Because Wi-Fi is generally limited to a range of about 100 m and cell towers can cover 1 km or more, Wi-Fi positioning generally yields more accurate results than Cell ID localization in areas with visible Wi-Fi routers. Cheng et al. from UC San Diego with colleagues from Intel and Microsoft (2005) showed that Wi-Fi positioning can achieve an accuracy of 13–40 m in urban areas with many SSIDs. With Wi-Fi positioning, the mobile device must first scan all available wireless access points in range. This gives the device a list of SSIDs that identify each visible network, along with relative signal strength to each access point. This data can then be used to query a server to determine the device's location, much in the same way that Cell ID location works. Because Wi-Fi must be activated and a sweep of SSIDs must be conducted, this method uses more power and is a bit slower than using Cell ID.

The most accurate location can be determined by using GPS or Assisted GPS (AGPS). However, this is also the slowest method and uses the most power. GPS satellites were first launched in 1989 to help boats and planes know their precise locations and plan routes to find their destinations and stay clear of other vehicles. Originally, only the government could access the full 10 m accuracy of GPS with civilian devices being crippled to receive only a 100 m accuracy in a system known as Selective Availability (GPS & Selective Availability Q&A 2010). This restriction was lifted in the United States in 2000 and the use of GPS for personal navigation and search took off. In order to obtain a GPS fix, a mobile device must have a direct line of sight to three different satellites and it must wait to receive signals from each one. Often, getting an initial location from a GPS system can take up to two or three minutes, as anyone who has ever seen the "Waiting for Satellites" message on a GPS navigation device knows well. Once a location is established, most modern devices can receive a new location every three seconds or so. However, running a GPS antenna draws a large amount of power. Tests performed by Amir (2010) on a recent HTC Android phone show that GPS adds about a 135 mA power drain when turned on. This is enough to drain a common 1300 mAh battery in about 6.1 hours taking into consideration the power needs of the processor and radio in the device (which Amir showed to be around 83 mA). However, since 10 m accuracy is needed for applications such as navigation, there are times when it is necessary to use GPS despite its large power draw. Applications that need a continuous GPS feed for hours at a time should be limited to locations where users have easy access to power, such as in a car. Likewise, applications that need only one precise fix should ensure that GPS is turned off after receiving a first accurate location to ensure that power is not significantly drained if users linger in their application for long periods of time.

Location on mobile devices is now a seamless part of the user experience of most new applications. Since the use of many novel mobile applications often relies on the user's context and environment, location can help almost any mobile user experience. Simple uses involve aiding search by limiting results to nearby venues, such as with movie or restaurant searching. More complex uses involve running location monitoring processes in the background and displaying currently relevant items in widgets on the idle screen, such as latest local weather or nearby events. Other location-based services provide information opportunistically, such as notifying users when their friends are nearby or delivering information to users in particular places as a form of digital graffiti where content is associated with a place.

These complex applications that continually run in the background can easily run into power constraints. Using lower-power methods such as Cell ID and polling location less frequently can increase battery life by orders of magnitude over running GPS for extended periods of time. Especially for situations that do not require precise location, these methods are critical to providing a user experience that does not leave the user with an empty battery halfway through the day but still can provide meaningful content based on their location.

For example, the Serendipitous Family Stories system (Bentley, Basapur, and Chowdhury 2011) defaults to network-based location and has two different intervals of checking location. When users are far away from the nearest story location, the device only checks for location every five minutes. However, when users approach a story location, the position of the device is checked every minute. This balance ensures that while users go about their day there is not a significant loss of battery life but that when near a story the user can receive notifications quickly as she approaches the area from which she can view it. In the continuation of this work for the public StoryPlace. me service, we refined the location algorithm even further so that it would default to GPS while a user is actively browsing nearby stories to give the most accurate results. As soon as a user leaves this screen, we default back to network location for our background process in order to conserve power.

By considering power, latency, and the required accuracy of position that is needed, the appropriate location technology can be chosen to provide the best user experience for each application. Choosing the wrong method can lead to long delays while users wait for their location via GPS or empty batteries that do not afford rich in-the-world experiences. The user's attention shifts then to the mobile device itself and why it is not providing the performance that is expected instead of remaining on their desired experience in the world (e.g., finding a friend, getting to dinner on time, exploring a city, etc.).

Data

Almost every mobile application today needs to connect to the Internet for some part of its functionality. However, using data on a mobile device requires a different approach than using data in a desktop PC application or on a website. For mobile users, data connections cannot be considered always available. As a user moves around throughout the day, their connection to the data network may come and go hundreds of times as they enter buildings, elevators, subways, or just sit in areas of weak connectivity. Even when connected to a data network, connection speeds will vary as users move in and out of 3G or 4G coverage or back to slower networks more reminiscent of dialup modems from the 1990s in terms of their data rates. This greatly impacts the average bitrate of the data connection that is available and can strongly impact services such as streaming video/audio or services that expect a persistent connection to a server such as chat or push email. Table 5.3 illustrates the theoretical maximum data rates on various types of networks, though practical speeds can be significantly slower. A test of network speeds by *Wired* magazine in 2009 found U.S. 3G networks ranging between 901 Kbps and 1940 Kbps compared to the theoretical limits of between 3.1 and 14.4 Mbps (Chen 2009).

Because of these changing data speeds, and because most users spend some part of their day in places without any network signal, it is important to design for changing

Table 5.3
Theoretical and observed data transfer speeds on various types of mobile networks. In practice, speeds can be 5–10 times slower in favorable conditions and might provide no throughput at crowded events such as baseball games. Observed speeds are from a class project where students downloaded a 1 MB file using major U.S. carriers and home/university Wi-Fi. Theoretical maximum speeds are from the published standards.

Technology	Uplink Speed (Mbps)	Downlink Speed (Mbps)	Observed Downlink Speeds (Mbps)
LTE (4G)	80	360	0.56–8.0
WiMAX (4G)	35	144	1.39–2.05
Wi-Fi	288.9	288.9	0.3–4.8
HSPA+ (3G)	22	56	0.47–1.8
HSDPA+ (3G)	5.76	14.4	0.03–1.36
EVDO Rev A (3G)	1.8	3.1	N/A
EDGE (2.5G)	0.9	1.9	0.011–0.136
GPRS (2G)	0.4	0.9	N/A

network conditions in new applications and in the larger systems in which the applications operate. It cannot be assumed that users have fast data connections just because their device registers a 3G/4G network or Wi-Fi. Therefore, it is important to always be conservative about the data that is sent to a mobile application, especially if the user needs to wait for that data to arrive before moving on to his next action.

Because of the Internet's reliance on persistent connections, the difference between standard Internet programming practices and mobile development is an important one. Over the years, our work at Motorola and in our class at MIT has required us to create applications that receive data pushed down from a server. For a desktop application, this is usually a simple matter of opening a persistent Transmission Control Protocol (TCP) connection to a server and using a protocol such as Extensible Messaging and Presence Protocol (XMPP) that will update the phone as resources change. While this basic pattern works on mobile devices, connections drop so frequently that it is necessary to be quite aggressive about reconnecting when data networks reappear. Also, because the device falls off of the channel and later reconnects, any messages sent over that time are not received. It is necessary then to ask for a history of updates that were missed in the time that the connection was not available. This adds up to quite a bit of additional data traffic (and power usage), especially if users are in and out of data coverage for several hours at a time. The resulting power drain might lead to a user's phone running out of power midway through the day. Also, many mobile devices turn off data connections that are not actively being used. If a device keeps a

TCP connection idle for more than a set amount of time (often 2–5 minutes), the data context may drop and with it the active connections. Sending keep-alive packets can force the connection to stay up, but with a resulting hit on battery life and processor usage.

There are several options to use for push connections that are available on the major mobile operating systems. Apple has a push channel that is available to developers (About Apple Push Notification Service 2010), and other platforms can use SMS messages to deliver content to a device that is otherwise sleeping to conserve energy. These options offer reliable delivery of information without the need to constantly re-query the server to find out what was missed, thus saving quite a bit of power. In the case of Apple's push channel, because the system uses one persistent channel for updates across all applications, it can save quite a bit of additional power on the device. The Android Cloud to Device Messaging Framework (C2DM) is a similar service available for Android devices that are running Android 2.2 or higher (2011).

Even with good connections, it is quite common that mobile data sessions will hang for 20–30 seconds in the middle of a download and then continue on. In an assignment for our class, almost all students observed this when trying to download a large file in multiple different locations. Mobile systems need to have long timeouts because a long delay with no data transferred might not indicate a failure. Latency in receiving the initial packets of data can also commonly rise into the tens of seconds. The user interface needs to respond to this with appropriate indications that data is being loaded so that users are aware of what is causing the unresponsiveness. This possibly long latency also demonstrates the need to properly cache data so that user interactions are responsive. In the Serendipitous Family Stories and StoryPlace.me projects (Bentley, Basapur, and Chowdhury 2011), we kept a cache of all stories and collections that a user was following on the device. With this cache, navigation within the application was responsive and did not result in additional queries to the server. Only when a user wanted to play a video was a server request required. This led to a much more fluid experience where users could find nearby content without waiting for a slow data connection to reload story lists and metadata from the server. Since receiving a serendipitous story is by definition something that removes the user a bit from their current actions, it needed to be as quick a distraction as possible.

On mobile networks, the cost in time of setting up a network connection is often much greater than it is over Wi-Fi networks or with web applications meant to be viewed on a traditional computer. Because of this, it is often a good idea to optimize network APIs in order to minimize the number of TCP sessions that must be created to support a given interaction. Instead of making three calls to a server in order to

Table 5.4
Theoretical latencies on mobile networks compared with data from the student project mentioned in table 5.3. In practice, students observed latencies from 230 ms to 10+ seconds on various networks from the time an HTTP request was made until the time the first byte arrived.

Technology	Latency	Observed Latency
GPRS	600 ms–1 s	N/A
EDGE	150 ms	401–4479 ms
3G (EV-DO)	120 ms	230–10396 ms
4G (LTE)	N/A	17–1633 ms

populate a list or render a screen, it is often more efficient to re-architect your system to make just one call. Although more data is returned, at the faster speeds of today's mobile networks the potential inefficiencies of sending slightly more data than is required in some situations will not normally be noticeable. The few extra seconds required for each additional network request to be created would be.

When users are on slow connections, however, excess data transfer can be noticed. It is often a good idea to compress the data stream that is being returned to the client, especially when sending large textual content. This can be accomplished by using less verbose representations of data such as JavaScript Object Notation (JSON) instead of Extensible Markup Language (XML) or by compressing the response through Zip or other standards, especially when the return data is large in size. These simple changes can make a mobile application much more responsive and let your users focus on their experience and their environment instead of waiting for excess data to download.

Finally, since users are often without a connection when they need access to information, caching becomes increasingly critical with mobile applications. Saving data and periodically updating caches makes sure that the freshest data is always available for a user, even if they need access to it while in a subway, on an airplane, or in an elevator. Applications like TripIt for Android and iPhone do a great job of this by storing all trip data between sessions and allowing full use of the application even when no data connection is present. In Serendipitous Family Stories (Bentley, Basapur, and Chowdhury 2011), we cache all story metadata on the device so that users can be notified of stories even if they do not have a reliable data connection. They can see the title and icon of stories even if they do not have sufficient bandwidth or signal strength to view the video. In this case, they can still unlock the story and can view the video later on when they enter an area with better reception.

Ensuring that data is available for users in a variety of conditions is an important aspect of mobile application development. Testing these conditions can be difficult, but is critical in ensuring a good experience. Developers should test applications by moving to areas with no signal while interacting with each feature of their application to ensure graceful handling of connection problems. Also, testing on a variety of networks helps to understand how the service performs in non-ideal situations. While 3G and 4G networks are starting to become pervasive, most phones still spend a considerable amount of time on older 2G and 2.5G networks. Applications should still be usable in these conditions. Users in many rural areas may rarely see a 3G connection on many networks. By appropriately pre-caching content or providing lower-bitrate files, systems can adapt to these slower networks.

Bluetooth

Mobile interactions need not occur just on the mobile device itself. With Bluetooth, companion devices can connect to the phone to move functionality onto the body or nearby environment. It is easier than ever to connect mobile devices to accessories with toolkits such as Kaufmann's (2010) Amarino, which enables Android Operating System (OS)–based mobile phones to connect to devices built with the Arduino hardware prototyping toolkit. Bluetooth modules are now fairly inexpensive and can be integrated into almost any device or article of clothing.

In our spring 2010 class, we explored various ways that mobile devices could connect to wearables in collaboration with Leah Buechley's New Textiles class at the MIT Media Lab (2010). The resulting projects illustrated a dozen novel ways in which our phones could inform us of information through our clothing or wake us up by softly vibrating our pillows. The possibilities for smart connected devices are nearly endless. Through Bluetooth connections to mobile platforms, clothing and wearable devices become Internet-connected devices with the potential for rich visual and tactile displays. The pillow project from our class connected a mobile phone to a pillow. In addition to softly vibrating the pillow to wake you up at the time set on your mobile phone, the pillow included LEDs that would glow different colors based on the expected weather that day. Another project was a bicycling jacket that was connected to a mobile phone. The user could input a destination on the phone, throw it in their bag, and start biking. Each time they needed to turn, the cuff on the appropriate sleeve would light up or vibrate, indicating the upcoming turn to the bicyclist. By connecting mobile phones to everyday objects, the objects become data-aware through the mobile device's Internet connection. This can enable a wide range of new experiences, but must be designed carefully to work within the constraints of the technology.

Table 5.5
Power consumption of the Bluetooth radio in various configurations.

Bluetooth Mode	Power (mA)
Discoverable	10
Connected	30

Source: Roving Networks, 2010.

When developing Bluetooth applications, many of the same guidelines from developing data applications apply. Bluetooth connections are prone to drop and reconnect frequently, so maintaining state and aggressively reconnecting when connections drop is necessary (Kasten and Langheinrich 2001). Bluetooth also uses quite a bit of power, especially when scanning for devices or while sending large amounts of data. Therefore, it may be advisable to cache data on the connected device or phone and sync periodically instead of trying to keep a constant connection open over long periods of time. Table 5.5 illustrates some of the power draw characteristics of Bluetooth that are useful in determining whether to open or disengage a Bluetooth connection.

When using Bluetooth, it is often the power use of the connected device, not the mobile phone, that limits connectivity. A shirt, wristband, or watch often does not have the same large-capacity battery that a mobile phone does. When designing an experience, it is important to take this into account. If a Bluetooth watch is going to serve as a display for incoming data from the phone but is kept in an unconnected state, the phone must first establish a connection and then send relevant data. While this might be acceptable for ambient background information such as updating the weather, traffic, or wellbeing data such as miles walked, it might not be responsive enough for real-time interactions such as phone calls. If it takes an average of two to three seconds to establish a connection, that might be the difference in a person being able to get to his phone before a call goes to voicemail or missing the call completely.

Mobile devices open up many new possibilities in terms of connectivity and interaction with the world. Connecting the Internet to devices worn on the body through a mobile phone can enable countless new experiences in the world that may be less interruptive and more ambient than needing to constantly check the screen of a mobile device for new information. Paying attention to power consumption and changing network conditions is critical in creating systems that need to run over longer periods of time. This can make the difference between a phone or connected

Bluetooth device running in sleep mode for days or running flat in five hours or less with GPS, Bluetooth, and the screen in use.

Environmental Capture

Today's mobile devices allow for easy access to the state of the environment around the user. Photo and video capture are the most obvious, but accelerometer, microphone, and compass data can all be used to provide new types of experiences on the mobile platform.

Andrew Campbell and his students at Dartmouth University have been exploring mobile sensing over the past several years. Through monitoring environmental audio they can determine when there is conversation occurring and through the use of accelerometer data they can determine when people are sitting, standing, walking, or running. This data is currently available in the CenseMe application and can be shared with friends on Facebook (Miluzzo et al. 2007). Explorations like these are only beginning to show us the power of the sensing capability of today's mobile phones.

Commercial systems are also taking advantage of these sensors. Popular applications such as Google Sky Map use the accelerometer and digital compass in order to determine the exact position and orientation of the phone and to display an augmented reality view of the night sky behind the device. Users can see planets and constellations as overlays as they move the device against the backdrop of the night sky.

However, there are technical constraints to consider when using this type of sensing. Any sensing that is continuous requires the processor to be running, which as shown earlier can be a significant power drain. Also, the sensors themselves use additional power. Amir (2010) has shown that the accelerometer on an HTC phone can use between 11 and 92 mA of power depending on the mode and refresh frequency used. This can be a significant power drain in applications that require a fast update frequency, such as games.

Mobile sensing and environmental capture are fast-growing areas of research and we are only beginning to see the possibilities of use when devices contain a rich variety of sensors. When designing new mobile experiences, we should consider how sensing can be a part of new applications in order to take the greatest advantage of the unique capabilities of mobile devices.

***

Each of these mobile technologies enables creating new types of mobile experiences. However, taking into account the limitations of each of these technologies is critical in creating applications and services that perform appropriately in a variety of conditions and do not overly drain the battery and leave users unable to make phone calls or send critical text messages. Appropriately managing power constraints remains the largest concern when running long-lived mobile applications and services. Meanwhile, developers and designers should ensure that applications respond appropriately when location or data networks are not available and still provide a meaningful experience to the user by using cached data or default content.

6 Mobile Interaction Design

Getting from a sketch of a concept to a full interface design can be a tough process. Mobile screens are small and it is almost never possible (or sometimes not desirable) to cram in all of the functionality you might be able to dream up. This chapter will take you through the process of designing user interfaces for mobile systems, from that one sentence concept that came from a generative research study to completed UI specifications and flows that match the needs of your users. The goal of this process is to detail a fun and enjoyable experience that will work in the context of daily life, keeping users returning to your application or service day after day.

Interaction design for mobile experiences is different from designing web interaction in two key ways. First, a mobile app is constructed for use in the real world, often in very short bursts of time and within an ever-changing context. Second, mobile devices have very small screens with relatively poor input mechanisms (usually a relatively large finger on a piece of glass). Like web design, however, interaction design for mobile experiences is the interface of many disciplines. Individuals who practice mobile interaction design typically come from a variety of backgrounds: information schools, design schools, human computer interaction, industrial design, or engineering.

This chapter will cover the steps of mobile interaction design: from idea, to conceptual model, to interaction models, to flows and screens. We have found that this process (like most work on mobile systems) works best in an iterative fashion where design flows back and forth between the concept and detailed levels of design and prototyping as the interaction is refined. Rapid prototyping tools such as Google's App Inventor (App Inventor for Android 2011) can help in quickly visualizing and exploring an interaction, but so can ordinary paper and pencil.

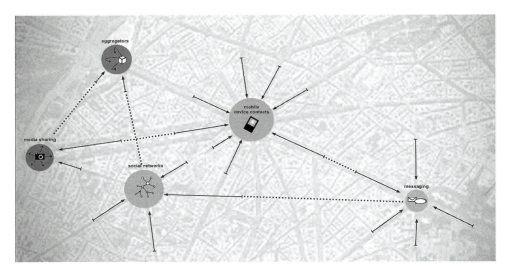

Figure 6.1
The mostly disconnected model of contacts on a mobile device pre-MotoBLUR. Multiple contact ecosystems each lived in their own siloed applications. Facebook contacts did not interact with phone or email contacts and it was not possible to link these identities together for contacts that existed in multiple places. (Image used with permission of Motorola Mobility)

Modeling

A high-level concept model is often the first step of a new mobile design, long before anything begins to be committed to a screen. A concept model helps to answer questions about how the experience will work in the lives of users and provides an understanding of the scope and benefits of your application/service. This helps focus more detailed design later on and aligns the ultimate design of screens and flows to work with a user's mental model of how he interacts with your application or service. Often, this can be the toughest part of the design, and is certainly the most abstract. Concept models are particularly critical for mobile systems, though, because less screen space can be devoted to explaining a system to users. Designers need to ensure that the system and the user are on the same page with regards to the operation and navigation of the app or service.

One way to create a concept model is to think about the world before and after your system is introduced. How does a user accomplish a certain task today? Or, how does a particular mobile ecosystem change after your solution is implemented? For example, students in our class created an alarm clock app which gathered data on

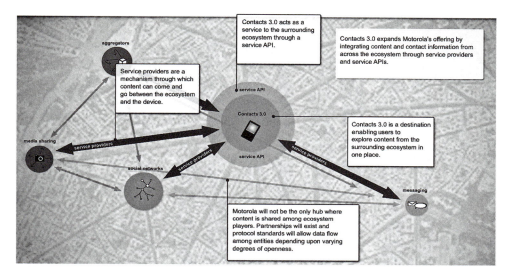

Figure 6.2
Contacts 3.0 allowed for aggregation of contacts from multiple identities into one place. All content would come to our middleware on the phone where we would present a unified view of a person's content/updates in the phonebook and allow communication via any means available. (Image used with permission of Motorola Mobility)

weather in order to alert users to a good powder day on the slopes, a day users would want to get up much earlier to make it to their favorite ski area. The students created a before-and-after concept model showing people constantly waking up to check the weather in the "before" while sleeping soundly until the optimal time to wake in the "after." This model helped the students to focus on the key value of their system, the added peace of mind users would have when utilizing their system. This helped in designing prompts and explanations of how the service was working to put users' minds at ease.

Another common way to create a concept model is to use a metaphor to express the core value of the system. At Motorola, Frank and colleagues (2010) used the metaphor of Parisian streets with a before and after concept model while working on the Contacts 3.0 service that would serve as the concept for the phone book in MotoBLUR. We noted that there were islands of activity in the contacts ecosystem on a mobile device, places where contacts converged. These were much like the layout of Paris with numerous dense nodes and hotspots of activity. We mapped out this model, which can be seen in figure 6.1. In the world of Contacts 3.0/MotoBLUR, we saw the roads between the nodes acting as a way to bring data together and merge information from multiple sources (much like the cars and buses traveling from node to node in the city). This model helped us to focus on the value of the platform that we were creating. We were to be the roads, bringing data together from multiple locations (hopefully a bit more efficiently than the traffic on Champs-Élysées!).

A conceptual model can also take other forms. It might be the user's model of the steps in a process, a look at the different tasks or mental models with which different user types approach the system, or a more software-architectural view of how the concept fits into existing systems and workflows. When designing Serendipitous Family Stories (Bentley, Basapur, and Chowdhury 2011), we created a model of places in people's lives and how users would interact with the system in various locations. For example, at home users might create videos or browse content that they have already discovered. While on the go, they might come across stories that are saved in the places through which they travel. While out in the city with extra time to spare, they might use the interface to find the closest story and walk over to explore it. These location-based concept models are a useful way to think about a system that allows for different interactions or uses in different locations in the world.

Another concept model that is often beneficial to create is a service blueprint. First developed by Shostack (1984), this model is especially helpful when designing services that users interact with over time or services where users need to interact with multiple entities that represent the service. For example, when users interact with an airline they book a plane reservation by interacting with a booking website. Then, on the day of the trip they access a mobile boarding pass, are notified of upgrades/delays, interact with staff at the airport and staff on the plane, and finally interact with the staff and environment of their destination. A service blueprint ensures that the messaging and interactions across all of these steps are consistent. The blueprint itself contains five components: Customer Actions, Onstage Actions (that the user can see), Backstage Actions (that are invisible to the user), Support Processes, and Physical Evidence (the artifacts that users can see). Service blueprints help designers to focus on the touch points where users and the service connect, ensuring that the entire experience, not just the screens on the mobile device, is consistent and enjoyable and that users have an accurate mental model of all of the parts that require their interaction.

The main objective of the modeling is to help the designer think about the goals of the system in new ways, so completing multiple conceptual models can often help in understanding the full scope of a new concept. Answers to these kinds of initial probes allow a designer to map functionality to various goals. Some of these aims might be simple tasks that a user needs to perform while others might be larger life goals such as connecting to friends or being a better spouse or parent, or simply having fun. The purpose of a concept model is to develop an understanding of the core value of your system from the user's perspective and prepare for the next step of exploring use cases.

Structure and Flow

How do we move from the abstract to the actual? What tasks must be performed, and how will the user navigate the interface?

This process is not a top-down, linear one, but rather moves back and forth between descriptions of use cases mirrored by possible wireframes (sketches of screens without graphical detail) in an effort to depict the flow of a user's experience. These user flows demonstrate users' possible movements through time: how they initiate a process, how they complete it, and what path they take. That flow demonstrates simple cause and effect, which shows the consistency of the system. This consistency inspires a visual language, one that is helpful and specific to the use of the product, a language in which there is no one "right" vocabulary.

The first step in making a design more concrete is often listing the use cases or tasks that a user of the system will perform. This is critical for mobile design as the screen real estate is small and all possible tasks may not be able to be supported cleanly. (There is even a subset of the design community that believes strongly in "building mobile first" (Wroblewski 2009)—that thinking about the tasks that are really necessary and what will fit on a mobile screen will result in a better full web experience because the hard prioritization decisions were made when the costs of complexity were highest.) A list of use cases can take many forms. When the design team is small and working together, it can simply be a list of phrases: tasks that the user will perform. Often, when presenting to larger audiences or getting buy-in from engineering or product teams, short descriptions or scenarios work best. However you choose to represent each use case, the representation should be a single user goal, something a user would turn to your system to do. Examples from Contacts 3.0 include "finding out what friends were doing," "commenting on a photo," and "responding to a Facebook message with a text message/phone call/email." After creating an exhaustive list of everything a user might want to do, some prioritization is almost always in order. Many of the use cases listed will be central to the experience that you are trying to create while others might be less so. We often have our students sort their use case list into three levels: Core,

Important, and Nice to Have. Other ratings such as MoCSoW (Must, Could, Should, Won't) can also be applied (Clegg and Barker 2004). These ratings can then be used when beginning to create the interaction framework and flows. Often, the design process will involve moving back and forth between this use-case list and the design in order to create a concept that is coherent and flows from one activity to another.

From a list of use cases, we move to an interaction model. The model is a first step toward developing the flow of an application: Which screens will be utilized and in which order will they be presented? How can a user get from one function to another? How do use cases combine to form tasks? These are all questions that an interaction model answers.

Because our class at MIT is only twelve weeks long, the experiences our students create are often simple with a handful of screens. Most commonly, their interaction model consists of linear, step-by-step, "wizard-style" flows for particular tasks. For example, setting an alarm clock to wake the user up if the snow on a particular mountain is at its optimal depth requires a few steps to pick the particular mountain, the desired wake-up time range, and options about how "optimal" the snow should be. A project to log workout data is equally linear in picking the type of exercise that is performed, setting the duration/weight/reps/etc., and seeing graphs of performance over time.

One much more complicated interaction model came with the Contacts 3.0 project at Motorola. Here, we had the challenge of exploring many different facets of the contacts in a mobile phone book. First, there was the contact list itself. There is always the simple A–Z list of contacts, which can be quite useful for directed tasks such as calling a friend. But we wanted to transform the phone book from a static list of names and numbers into a destination that people visited in their down time to explore what was going on with the people in their lives. This led to several other views of the list of contacts, including a view where users could see the people that they have communicated with recently (through phone calls, texting, email, social network messages, etc.) and another view where they could see the latest updates from their contacts across a variety of services such as Facebook and Twitter. An interface where users could swipe between these views seemed to fit these use cases nicely.

That interaction model worked well for the list of contacts, but we faced a similar issue when designing the detail screen for a particular contact. There was too much information to put in a single view, so we adopted a similar swipe design where users have one view to see the traditional phone book details (name, phone numbers, email addresses, physical addresses, etc.) and other views for recent communication with that contact and her recent social networking updates. We saw the potential for many

Communicating — Priority

Use Case	Priority
Call a Contact	C
Send Message (or reply to) a Contact (SMS/MMS)	C
Send Message a Group (SMS/MMS)	C
Email a Contact	I
Email a Group	I
Send Lightweight Communication to Contact	N
Send Lightweight Communication to Group	N
Instant Message a Contact	N
Instant Message a Group	N
Send Voice note to a Contact	N
Send Voice note to a Group	N
Post on someone's Wall on Facebook (or equivalent for another Service)	N
Send Message to a Contact through an online service (i.e. Facebook)	C
Request info from a contact (Location, Contact Info, etc.)	I
Comment on Contact's media	C

Learning

Use Case	Priority
View Contact's Status Message from online services (i.e. Facebook)	C
View Contact's Mood	N
View Contact's exact Location (cross-streets, address, dot on map)	I
View Contact's vague Location (City, State)	I
View Contact's user-defined Location (home, store, tag, etc.)	I
View Contact's time zone	I
View Contact's distance from me (exact - .2 mi)	I
View Contact's Motion Presence (moving/not moving; duration)	I
View Contact's distance from me (general - same city, near, etc.)	I
View Contact's preferred communication method	I
View Contacts Music Status	I
View Contact's current/recent photos	C
View Contact's comment on a photo	I
View Contact's recent posts/online activities (short: eg Twitter)	C
View Contact's recent posts/online activities (long: eg. Blog)	C
View Contact's comment on a Blog post	I
View Contact's Facebook Profile update	I
View Communication History with Contact (Recent Calls, etc.)	C
View Business specific data (Hours, Locations, etc.)	I
View Contact's Calendar availability	I
View Contact's IM Status now	N
View Contact's Calendar	N
View Contacts contacts, network	N
View Contact's online profiles (Facebook, Myspace, LinkedIn)	C
View Contact's weather where they are (attach Widget to a Contact?)	I
View Contact's Ring Profile (Ring, Vibrate, etc.)	N
View basic Contact Information (Phone number, Email, IM, etc.)	C

Sharing — Priority

(From My Info section on the Social Dashboard)

Use Case	Priority
Set my Status (make avail. OR push)	C
Set my photo (to appear in others' Contact lists)	C
Set my Mood (make avail. OR push)	N
Share (make avail. OR push) my exact Location	I
Share (make avail. OR push) my user-generated Location	I
Share (make avail.) my Motion Presence	I
Share (make avail. OR push) my vague Location (City, State)	I
Share (make avail. OR push) my Time Zone	I
Share (make avail. OR push) my Calendar	I
Share (manage permissions for) my current media (what I'm watching, Blogs, Music?)	C
Share my preferred communication method	N
Share (make avail.) my IM Status	I
Share my online identities (url to any online profiles that a user has)	C
Share myself as a Contact (my Vcard)	C
Share my Contacts/my Network	C
Make an Introduction to a Contact (like LinkedIn) (sending Contact Info (like sending Vcard)	I
Send content to a contact (link to media, Blogpost, etc.)	C
Send content to a group (link to media, Blogpost, etc.)	C
Send Meeting/Event Invite	I
Share business-specific data (if Contact is a business)	N

Searching/Finding/Browsing — Priority

Use Case	Priority
Search for a Contact on-Device	C
Search for a Contact off-Device	C
Search by Tag	
Browse Contacts	C
Filter/sort Contacts Main by:	C
Tag	I
Location	I
Now (or recently) Playing [media]	I
Recency of communication	C
Recent Updates	
Calendar (Upcoming)	
Franchise Specific activity (Now playing media)	
Group	C
Favorites (suggested by frequency)	C
Filter/sort Social Dashboard Content (Device + Social/Web updates) by:	C
Group (from Contacts App)	C
Media Type (off-Device)	C
Recency (time-based by default)	C
Communication Type (on-Device: Calls, Messages)	C
Web (off-Device)	C
Phone (all on-Device)	
Web Service	C

Creating

Use Case	Priority
Create a Group (from Tags)	C
Create a Group from a shared group (a Contact send/shares the Group with another member)	C
Create a Group from a Message (sending to multiple recipients)	N
Publish Group (notify) Contact that they are in a Group and give option to add group)	I
Create a Group from scratch	C
Create a Contact from Web Service (custom API for Facebook, etc.)	C
Create a Contact from another Device (including "Kissing" and tagging (location))	N
Create a Contact from scratch	C
Create a Contact from Structured Data	C
Request an Introduction to a Contact (requesting Contact info for another person)	I
Rate Contacts (make Favorite)	N
Block a Contact	C
Block particular feeds for a Contact	C
Block a Group	C
Disband a Group	C

Syncing/ Backing up/Storing

Use Case	Priority
Sync/update Contacts with Web Service	C
Sync/update Contacts with another Device (PC or second phone)	I
Save Contacts to SIM	C
Save Contacts to Phone	C
Save Contacts to Service (NGP - back-up?, carrier?, 3rd party?)	I
Customize/set preferences for Contacts Detail	C
Customize/set preferences for Contacts Main	C

Setting up

Use Case	Priority
Import Contacts from Online Service	C
Reconcile/merge Contacts across information sources (matching John on Facebook to John in Contact List)	C
Set which content sources I want from a given Contact	I
Receive notification that a Contact has edited their Contact Info and there is a conflict (following Auto-Sync)	I
Respond to conflict alerts following automatic sync/updates with online services (in background)	
Set sync (Web services) option to manual	
Configure Automatic Sync	C
Choose whom to add from a given Service	C

Figure 6.3

A partial use case list for Contacts 3.0. Use cases were grouped by larger functional areas and then rated as (C)ore, (I)mportant, or (N)ice to Have. (Image used with permission of Motorola Mobility) views here including a depiction of the weather and time zone where the contact is currently located or a view of their calendar for the current time. Our initial interaction model can be seen in figure 6.4. This model can be considered a matrix, with interaction possible in multiple dimensions at a time. Matrix interactions are more complex and it must be clear to users where they are in the interface and where they can go at all times. Matrix interfaces can also confuse the concept of "back," as switching between views on one screen may or may not be seen by the user as transitioning to a new screen in the interface. However, when information presentation needs are complex, matrix navigation paradigms are often necessary and can greatly simplify the interaction.

Another common mobile interaction model is a hierarchy, where users are presented with a list of options, dive into a particular option, perform a task, and then "back" out to the main menu or another higher level in the view hierarchy. The most famous example of this interaction style is the iPod, with hierarchical menus for delving into artists, albums, and songs with frequent use of the back button to change to other selections. This interaction style can be seen in many applications on the market today. In a way, our Contacts 3.0 design was also hierarchical as users could go from the list of contacts into a particular contact and then perform some action on that contact (call, comment on a Facebook status, etc.). While creating an interaction model, it is often common to go back and forth to the list of use cases, exploring how various use cases link together into tasks and how the interaction model could support these in more efficient or intuitive ways. While working through an interaction design, the priorities of use cases might also change as designers continue to develop a stronger understanding of the core values of the new application or service.

Designing the Screen: Interface Design Principles

Designing for a screen—a very small screen—cannot be reduced to a set of instructions. But certain basic principles make sense. For example, Wodtke's (2002) eight principles, while originally designed for the desktop web, can help in making sense of potentially complex interactions on the mobile device. Designing for "way-finding," a visual scheme that signposts where you are, where you can go, and how to get there, can help ensure that users do not get lost. Mobile interfaces often do not afford much screen real estate to navigation controls like a website usually does. Making sure users do not get too deep in an array of screens that they will have to back out of one at a time as well as making it easy to get "home" to the landing screen of the application

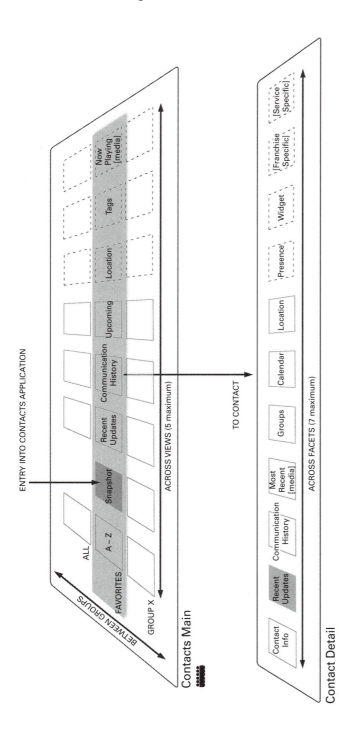

Figure 6.4

The initial interaction model for Contacts 3.0. Both at a list and an individual contact level, users could swipe back and forth to see different views of their contacts. This helped turn the contact list on the phone from a simple list of names and numbers to a rich destination that could be explored to learn more about the people in one's life. (Image used with permission of Motorola Mobility) are good principles for mobile design. It is also important to take into account physical limitations of a handheld device (small screen, small icons, the changing optic scape of users as they age, how hard it is to use a touch screen with gloves on when you are outside in wintry weather). Buttons that you might think are large enough might not be for someone riding a bus with his phone constantly moving up and down. Also, different phones have different pixel densities. If you are designing in absolute pixels, it is important to understand the physical size of the graphical elements on different popular handsets to ensure that they are still large enough to interact with or read.

All of this is easy to say, hard to do. Interaction flows and screen design objectives are simply good, clear thinking written down. But arriving at these design moments requires an iterative process open to a range of inputs.

Interaction flows do not need to be carefully illustrated documents. Creating them is often a process of much iteration. In class, we have our students create their initial flows with pencil and paper and work through several iterations before they are ready to build their first paper prototype. We combine the design with several rounds of paper prototype testing as described in the following chapter. Most of this design iteration occurs during a three-hour class session after which each team has a week to finish the completed first-draft paper design of its application and conduct additional usability testing.

This style works outside the classroom as well. In 2006, Frank and Vernell Chapman at Motorola were working with the Yahoo! Research Berkeley team to create a J2ME version of the ZoneTag application (Naaman and Nair 2008). This application had to integrate with the camera so that users could take photos, apply tags to them based on their location, and upload them to the photo sharing service Flickr. A seemingly simple task until one looks at all of the additional steps of provisioning the application with a user's login credentials, viewing uploads that are occurring in the background, and providing an interface to browse and select from possibly hundreds of suggested tags. We worked through this interaction several times and our final flow can be seen in figure 6.5. The pencil drawing might not be pretty but we were satisfied with the interaction and ease of tagging a new photo and moved on to implementation based on this flow. It was simply as detailed as we needed to begin implementation and share with other stakeholders. Not every design needs beautifully rendered wireframes in Illustrator or InDesign.

From flows, you can design screen level details. We have found that it is best to stick with platform conventions on most devices. If there is a dedicated back button, this should be used. If most apps place navigation controls in particular locations on

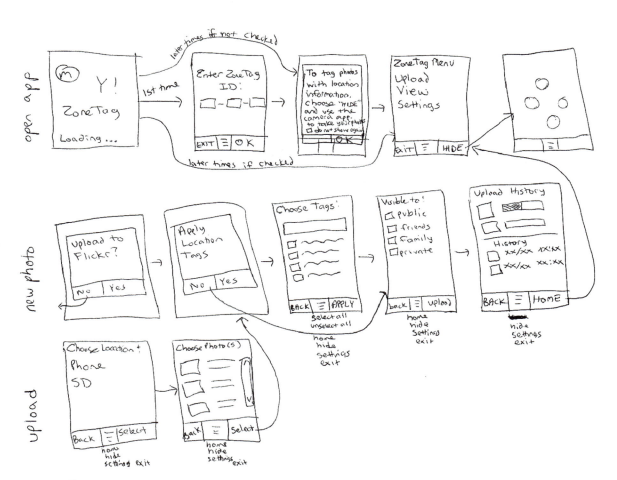

Figure 6.5

The initial user flow for the J2ME client for ZoneTag. With the need to set up the application, upload photos with location-based tags, view uploads in progress, view nearby photos based on location or tag, and offer settings, the application required a large number of screens. We were able to fit these into an easy-to-use flow that is hierarchical (starts with a top-level menu) and linear (taking a photo or browsing photos is a wizard-like experience). (Image used with permission of Motorola Mobility)

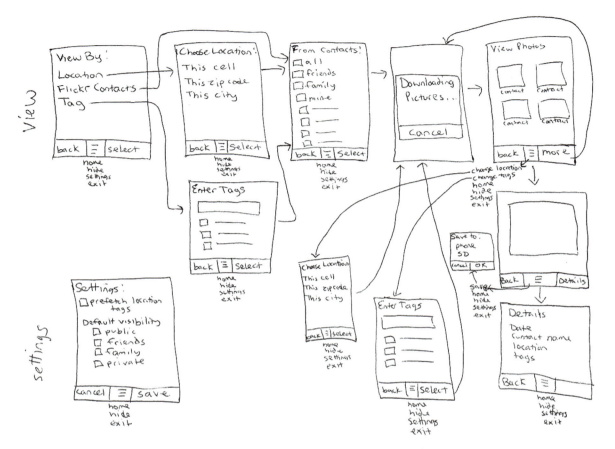

Figure 6.5
(Continued)

the screen, it is generally best to follow these patterns. Users expect a consistent experience across all applications on their phone. This often means that particulars of an interaction or screen design will need to be created multiple times for multiple different platforms. When we created the TuVista service (Bentley and Groble 2009), we created Android, iPhone, and Blackberry applications and each corresponded to the platform's conventions as can be seen in figure 6.6.

Designers should also always give users feedback when potentially long operations are occurring. Each platform has its own standard way to do this, which should be followed. Spinning icons, for example, can inform users that a networking operation

Figure 6.6

The TuVista interface on iPhone, Android, and Blackberry. The iPhone design has the dedicated back button and tabs on the bottom while the Android application relies on the hard key back button and has tabs along the top. Tab styles correspond to the platform conventions on each operating system so that users are presented with a style that is familiar and understandable. (Image used with permission of Motorola Mobility) is pending. As a rule, assume that any network operation might take ten seconds, as this type of response time frequently occurs in practice.

The ultimate goal of any interaction design project is to make something that is usable for a wide variety of potential users. This is a process of much trial and error. The following chapter will discuss the process of usability evaluation. This often occurs in a repeated cycle with flow and screen design and it is important to keep iterating until you come across a design that works for people and is easy to understand and navigate.

7   Usability Evaluation

In chapter 2 we discussed a user-centered design process as applied to mobile apps, a process we described as predominantly qualitative in approach and highly iterative. But to say that the "design process" is a distinct and separate initial stage in application development would be incorrect. "Design" might better be thought of as a continuing feedback loop of prototypes in various stages of development. A critical stage of that development is usability evaluation.

"Usability" is certainly a widely established concept in web development, largely owing to the work of Nielsen (2003; Nielsen and Loranger 2006) and Norman (1990; 1998; Norman and Draper 1986) in the 1980s and '90s. But accessing a website from a computer is still an activity that takes place predominantly within a set locale (office, home, coffee shops, and other Wi-Fi hotspots), using a device that, while transportable, is still not as portable as a mobile phone. Complicating Nielsen's very useful and widely adopted principles of usability is the remarkably fluid context in which mobile apps are used. That context includes the physical dimensions of the mobile device (small screen size and keyboarding elements); the choice of carriers all with differing characteristics and strengths (and weaknesses); cost differentials among data plans; the ever-changing real-world setting in which we invoke mobile apps; and, importantly, the implicit demands and expectations a human being projects onto a mobile device, the unconscious needs and expectations that may be difficult to recognize or articulate until confronted with a real-world, immediate situation.

These real world situations, and how computing devices can be used in them, is the focus of a large international research agenda: ubiquitous computing. At the start of the movement, ubiquitous computing meant living in an environment with computing integrated into ordinary objects: refrigerators, toasters, cars, and clothing. When we first came to MIT, graduate students in the Media Lab such as Thad Starner (Starner et al. 1997) were experimenting with early "wearables." These students trekked around campus with bulky backpacks containing a Central Processing Unit (CPU), wearing eyeglass frames with a small camera attached for media capture and a tiny monitor screen covering one eye so that they could follow various computationally mediated interactions. They managed to reduce in size all the elements we associated with networked computing. The gear and the appearance of a person bedecked with all the gear may have engendered a range of reactions from excitement (if this was your thing) to amusement or scorn (a typical response to some early digital mania). But this research agenda offered an early take on the notion that "ubiquitous" could also mean something other than embedding computation into things we already own and places through which we navigate in our daily lives. As Savio and Braiterman (2007) in "Design Sketch: The Context of Mobile Interaction," say, it is no longer sufficient to talk about mobile devices; it is necessary to talk about a new concept of "interface" as the boundaries where device, person, and phases of use intersect. As they write: "We must focus our design practices around mobile people, not mobile devices."

In other words, we are not merely shrinking in size a Web experience, but creating an entirely new platform for communication and interaction. And this "new platform" is life as it is lived: private, social, and irreducible to formulaic expression. As Nakhimovsky, Eckles, and Riegelsberger put it in their CHI 2009 workshop proposal on "Mobile User Experience Research":

Interaction with mobile devices, as the most mainstream manifestation of ubiquitous computing, is a site for the collision and combination of different approaches and theoretical concerns in human-computer interaction (HCI) research.

And this interaction with mobile devices that are inextricably connected to the larger context of real-world use requires a wide range of methods for creation and evaluation. How then to evaluate mobile applications that purport to be complicit in our negotiation with the world around us, with our lived experience, as direct, indistinct, and partially understood and articulated as it always is?

In a study of Google Maps for Mobile, Riegelsberger and Nakhimovsky (2008) rightly point out that some of the methods in mobile user experience (UX) research have their limitations, a statement widely recognized in the field of mobile research and development. The "variable context" of use in mobile apps becomes a limiting factor in evaluation, a factor that cannot be fully addressed in lab evaluations that web developers might traditionally use. Riegelsberger and Nakhimovsky's field study reporting on user experiences of Google Maps for Mobile—an experience that "results from the interplay of carriers, handset manufacturers, and application providers . . .

a mobile ecosystem"—highlights the need for collaboration between users and product developers in the context of real interaction in the world.

Their study suggests that valuable information on user experience results from a blend of the methods we discussed in chapter 2. Especially helpful in their view were quick one-to-one participant debriefs of initial field tests, followed by product team debriefs to define quick fixes as action items. These quick initial participant debriefs allowed developers to zero in on small usability issues that might go unnoticed in a lab setting but that assume great importance in real-world contexts. Equally important, the researchers report a valuable ancillary effect of this early collaboration of user and developer: users became more vocal about small inefficiencies as testing continued. Coupled with ongoing informal telephone interviews, users became more aware of these unmet, implicit needs, and their comments became a basis for follow-up product development brainstorming.

For Riegelsberger and Nakhimovsky these field tests and debriefing loops were conducive to substantive product development data. They were convinced that a supple evaluation method, not tied to established protocols, aided development of a system. One-to-one debriefings between a developer and a user and later group debriefings with all users in the study, combined with usage data derived from diary studies and logging methods we discussed in chapter 3, created what they term a "co-inventor relationship" with developers. This relationship, they concluded, was beneficial to the production team for its quick identification of technical insufficiencies (rather than waiting until the end of a formal study when "small" matters may be overwhelmed with data) and its recognition of "varying interpretations of observations," which led to a shared understanding between researchers and users of the successes and failures of the system and, therefore, was important for the next iteration in the development cycle (Riegelsberger and Nakhimovsky 2008).

Lab Usability Evaluations

Because mobile experiences occur in everyday contexts, the concept of a lab usability evaluation does not translate as well as it traditionally has for web applications. For most people, using a website on a computer is an activity that is undertaken in a single place for a set duration of time. And most computers can be approximated to be the same with a large screen, keyboard, and mouse/trackpad. Therefore, a computer in a lab can adequately substitute for a computer in the home or work. It is still uncommon to see people on a street corner take out a laptop and look something up for a few seconds. This is the unique domain of the mobile device.

Mobile usability, then, is more than just the ability to navigate from screen to screen and understand the prompts, icons, and flow of an application. That is not to say that the understanding of these concepts is unimportant; they are necessary but not sufficient. Quick and informal lab usability studies can still be used to validate prompts, icons, and flow, and these studies can be run much in the same way as a traditional web usability study. Following Nielsen's (2000) findings, a small number of users (usually five to ten) can be brought in and asked to perform some simple tasks while thinking aloud as researchers note areas for improvement. We recommend to our students that this step be performed at a very early stage, usually while the design is still in paper form. Paper allows quick changes to be made and studies have shown that participants in usability evaluations are more critical and identify more problems when using a paper prototype. This is because they do not feel that their comments will necessitate a large amount of rework since nothing is built or designed to a high level of detail. Also, Rettig (1994) points out that if a prototype is made to look very high fidelity, users will comment more on the "fit and finish" issues than on the overall interaction and understanding of flow, which are the most critical topics to get right at this stage of design. Doing this evaluation early helps to ensure that the initial interaction model makes sense to potential users.

If testing on a functional prototype, some things can be done to achieve more real-world results. Often, mobile usability tests are conducted in a lab where mobile networks are strong or Wi-Fi is being used. This can make the resulting mobile experience load quickly and results in a scenario more ideal than typical. Testing applications in both good and bad network conditions can help the research and design team better understand points of user frustration and identify data that should probably be cached or pre-fetched to improve the flow of the experience.

Observing a mobile usability test can also be a challenge. There are many tools available today to capture web interactions. From commercial software such as Camtasia (2011) that captures all interaction on the screen to more complex eye-tracking solutions and the use of JavaScript to log mouse positions, there is a large toolbox from which to choose (Jacob and Karn 2003). Capturing interaction with the mobile experience is much more difficult. Researchers at Google have built camera mounts that suspend a camera above the screen to capture the interaction. While a good initial step, these cameras can get in the way of how users hold devices, and occlusion from a hand between the camera and screen can limit researchers' ability to see all of what is happening in the interface or exactly where a person is touching.

Some research projects on the Android platform (Balagtas-Fernandez and Hussmann 2009; Taking Screenshots . . . 2011) have built tools that periodically capture screen shots on the device; however, it is quite easy for these systems to miss the exact location of a touch or what was happening on the screen during a transition. Tethering a mobile device to a TV or computer using High-Definition Multimedia Interface (HDMI) cables is another solution that is possible on many of today's mobile phones (Real HDMI 2011). While this may interfere a bit with how the device can be held and rotated, it is currently the best way to capture live video of mobile interactions for recording and later study by the research team.

Lab usability evaluations can also be quite a useful tool for testing non–context dependent interactions. When Frank and colleagues (2010) at Motorola were developing the social networking aggregation for Contacts 3.0, we had questions about the ability of a simple phone-based algorithm to correctly match contacts together. For example, if a user connected Facebook, Google, and Twitter accounts together, could the system accurately merge a person's contacts together when they had identities on multiple systems? We also had questions about the way we were proposing to present this matching process to users, especially around matches we were not totally sure of (e.g., is the Jess Smith in a user's phone book the same as Jessica Smith on Facebook with no other matching information?). To test the matching, we performed a lab usability test one day in our Chicago office. We recruited ten people to come in throughout the day to use the system to add their various accounts. We copied the contacts from their SIM cards to our phone and observed as they added Facebook, Flickr, and/or Last.fm accounts using an early prototype of our system. Through this simple one-day evaluation, we observed many possible issues in aggregating contacts using real-world data and users. We were able to use these results to make improvements to our design for the product concept.

These lab usability tests are a good way to ensure that an application or particularly troublesome use case is in fact usable before embarking on a larger field study. They are quick to execute and gather the top usability issues with little effort and just a few users. However, as mentioned earlier, the issues that can be measured with a lab usability test (mainly issues in prompts, icons, and flow) are just a small part of the usability of a mobile application or service. Understanding how usable an application is in the places and contexts where a user actually wants to use it is a very different matter that requires different methods. If a user cannot get the information she needs in the ten or twenty seconds she has while waiting for a bus or train, your system might not be the first one she turns to. Or if the phone keeps going to sleep while a user is in the middle of another task such as cooking, he might not have a clean hand to wake it back up and find the next step in a recipe. Understanding these usability constraints and how a new application complies with these constraints requires some field evaluation, as will be discussed in the following chapter.

Our experiences teaching an MIT class in building mobile applications certainly reinforces Riegelsberger and Nakhimovsky's (2008) conclusions arguing for a more supple method of evaluation. We employ a process of rapid prototyping and iterative analysis and evaluation (see chapters 4 and 8) that relies on real-world contextual data gathering and frequent group collaborations to discuss and tease out meaningful interaction issues and improvements. A twelve-week semester for a class that meets three hours once a week imposes a tight timeframe in which to take an idea (proposed by development teams of three to four students each) from inception to completion. All stages of application development tend to melt into each other given the fore-shortened, hothouse environment of an academic term. But this abbreviated development cycle offers important insights into the idea of usability testing as an ongoing, collaborative process: a collaboration not only among members of a development team, but also among developers and potential users.

For example, our class usually consists of fifty to a hundred students, none of whom have worked together before the first class meeting. Once our students have thought about the design of their applications and services, mostly following the methods outlined in the earlier chapter on design, they create a paper prototype of their system. Each time a team presents its prototype in the studio setting of class or gives an Android Application Package File (APK) to another team, informal usability testing is occurring. We have found that iterating rapidly with multiple people reviewing each interaction helps to quickly improve the understandability and usability of the resulting system.

Paper Prototypes

A paper prototype is an early version of a concept design where screens are drawn on paper for users to interact with. Because it is on paper, it allows for easy editing without much wasted effort. If some part of the interface is not working for people, it is just a few seconds of work to redraw it in a way that is easier for people to understand and use. Paper prototyping was best promoted by Marc Rettig in a 1994 article in *Communications of the Association for Computing Machinery*. In this article, he discusses the benefits of paper prototyping as well as how a usability session should be run.

Following many of Rettig's guidelines, and adopting a few that are more suited for mobile development, we have our students create their prototypes with each screen on a sheet of paper. Dialogues and menus are then cut out to represent any overlays.

Torn out pieces of post-it notes can be placed into text fields so that each user can make his/her own entries without destroying the screen for future users.

We ask students to prepare three tasks that correspond to their most frequent use cases or use cases where they expect interaction issues and need the most feedback from users. We ensure that they describe these tasks in neutral terms—not referring to words used in the interface itself to avoid giving users hints about where to explore. Project teams—now representing potential users—circulate around the classroom "using" the app as represented in the paper prototype. It may be obvious but important to note that now the "classroom" has been turned into another environment. No longer a top-down, instructional, presentational space, the classroom transforms into a busy—and messy—agora, a testing space for ideas and approaches. Students try out these tentative designs, asking questions and probing why something was done a certain way. Project developers are cautioned not to be defensive about reactions from users, particularly when a user cannot understand why a particular sequence of actions or screens are the way they are. These are the crisis points of creativity, and developers should be silent in the face of them, asking questions only when interactions are complete.

The mechanics of this process are fairly straightforward and decidedly not high tech. As shown in the accompanying figures, students are excited about the opportunity both to get feedback on their own designs and to try out other groups' projects. Some of the prototypes are quite low-fidelity pencil sketches while some groups create realistic renderings of their screens on the target platform.

To conduct the evaluation, one student in the development team acts as an impassive "mobile app," presenting different screens in response to a user from another team's actions in touching the paper "interface." Another development team member hands descriptions of various tasks, each on a separate index card, to the user. Remaining development team members take notes on a user's performance, and record comments from the user's think-aloud responses (see the discussion of Task Analysis in chapter 3).

In the space of two hours this procedure is repeated many times per project. At the conclusion, each development team has a rich collection of user responses to a tentative, sketched application prototype. Some general guidelines are key to the success of this process:

• Screens are not overly designed—simple pencil on paper "screen" designs are used, without color or other eye-candy considerations

• Testers are instructed in think-aloud protocols before the test

Figure 7.1

Paper prototypes can be extremely simple pencil-sketches as shown on the left or can look more like the final solution using platform-specific UI elements. Research has shown that users feel free to be more critical of lower-fidelity prototypes and these serve as an excellent vehicle to get very early-stage data about the usability of a new concept. Photos by Audubon Dougherty. Used with permission and permission from students.

Figure 7.2
Students conduct a paper prototype usability evaluation. One student acts as the "computer" preparing screens and dialogues while another takes notes. A third, non-team member, performs the tasks asked of him. Photo by Audubon Dougherty. Used with permission and permission from students.

- Observers only observe and take notes on what is happening—no Q&A with testers during the test is permitted (but can and should follow the interaction session)
- Users are debriefed immediately after the test and all comments are recorded; Q&A between developers and testers takes place now
- All groups are required to summarize and report on their observed results, orally at the end of this class session and more fully in a short written report due the following class

This last guideline also requires each group to refine their prototype in response to user evaluations. The time lag between classes allows teams to consider design refinements in response to comments they have collected from users—with the understanding that this next design, still on paper, is equally provisional and will go through another similar round of user evaluations. Of course, the semester is a limited, well-defined span of time, with a determined end point when student teams must produce a more or less completed app before their final project demo.

Outside the scope of the class, running a good usability evaluation generally requires between five and seven users. Tasks given to users should cover most major use cases of your system and especially ones that you expect will be confusing or troublesome. In general, facilitators should not interfere with the participant's task (some facilitators even leave the room), but there should be a clear failure point defined such that if a participant does not complete a task in that amount of time, the task will end. This keeps participants from becoming overly frustrated or angry after trying to attack a problem for many minutes without success. Failures generally point to larger usability issues that should be addressed immediately. By having participants "think aloud" while interacting with the prototype, researchers should be able to understand what they were looking for when they got lost and better design prompts, icons, or the flow of the system to align with users' own mental models of the system.

One year we experimented with bringing project teams into MIT's usability lab to test their paper prototypes. This lab is similar to a testing lab in a corporate development environment: two rooms, one for a tester and a facilitator who presents the tester with certain tasks to perform and another in which product developers follow the tester's screen interactions by video feed and listen to the tester's think-aloud descriptions of actions he or she is taking. Developers were asked to take notes on their observations, followed by a group debriefing of the test subject.

As we have noted before, this testing set-up could not possibly mimic the real-world context of use as a factor in evaluation. Probably the best example of the limits of lab usability testing of a mobile device was presented by a team developing an app that would enable car drivers to register road hazards on the warren of Boston-Cambridge city streets. The usability lab provided a distraction-free setting for evaluating the UI without the real-world hazards of Boston-Cambridge driving while accessing a mobile device (which is illegal in Massachusetts). The algorithm that the students had developed to detect a pothole while driving obviously could not work in this setting. The team created a wonderful toolkit for capturing, tagging, mapping, and communicating road hazards—but this experience was one that could not be tested in a realistic way in a lab setting.

Usability is an important part of mobile development. But it is only a first step. While usability testing in a lab setting can help validate initial interaction designs and conceptual models, it only goes so far in testing the fitness of a given design in real-world settings. Obtaining usability data from real-world field evaluations is critical and will be the focus of the following chapter.

8 Field Testing[1]

Early iterations of new concepts are best understood when they can be taken out into the world and used as a part of daily life. We have found that nothing matches the feedback from real people using a new mobile application or service in helping us understand how a given concept will integrate into the spatially and socially complex interactions of the world. Taking a new concept into the field is also the easiest way to understand where a current concept lacks needed functionality or suffers from more complex usability issues relating to context. Understanding the ways in which a new concept is appropriated by users and integrated into their lives helps in planning future versions of the concept as well as in preparing for marketing the product if it is heading for a commercial release.

These field trials are also used as risk-mitigation tools in order to understand how a concept fits (or does not fit) into real people's lives and interactions with each other and their data. They are not meant to address "usability" per se but rather to focus on the usefulness of a concept in everyday life as well as how and where it is used. Of course, usability issues will arise, and they are interesting to note and improve upon if a second iteration is made, but the focus of this stage of a project should be on evaluating the concept itself.

Once we have a prototype created, we have a number of different methods that we use to evaluate the experience. While these methods vary based on what we are testing and our research questions, the choice of our combination of methods is always based on a single guiding principle: we want the users to engage with the prototype in the course of their daily lives. Ecological validity, to us, does not mean approximating the real-life situation; we want the materials and settings to be the real-life situation. There are so many contextual variables that affect the mobile experience—where the person

1. This chapter is expanded from a short paper written with Crysta Metcalf for a workshop on Mobile User Experience Methods at CHI 2009 (Bentley and Metcalf 2009b).

is located, who else is around, their current needs and emotional states, just to name a few. Putting a prototype into users' hands for several weeks gives us a good indication of how people will actually use and react to the concept. Testing in a lab, or with paper prototypes, will not get us this rich interaction information.

Because of this guiding principle, we do four things in our research that we consider the basic criteria for testing prototypes. First, we recruit social groups (people who already know each other) when we are testing social technologies instead of asking strangers to act as if they know each other. Second, we put the technology in the field: we ask people to take it home, to work, and all of the places in between and use it as they would if they were not in a study. Third, we make sure that the participant needs to carry only one mobile device. And finally, we select data collection techniques that allow us to come as close as possible to "being there."

The remainder of this chapter will discuss each of these principles using real-life examples from work at Motorola and elsewhere. We will conclude with a look at discount methods to speed up the process of field testing while still getting useful results in the product development process.

Social Groups for Social Tech

Many of the applications we have created have been social applications—applications that were created to enhance experiences for people who already know each other. Because of this, we have found that it is essential to recruit participants who are friends or members of the same family. It is the knowledge of, and interest in, the other people's lives that makes the experience meaningful. For example, in our Motion Presence (Bentley and Metcalf 2007) study it was knowledge of her husband's schedule that allowed one of our participants to interpret that "moving" meant her husband was not yet in his meeting and thus was available for communication. Likewise, another participant in the study felt bad for his friend who was out and moving at an early hour of the morning on a work day, not previously knowing that his friend left home before 7am every day. If strangers are recruited to evaluate technology that is fundamentally social in nature, not only will they not be motivated to be social with the other participants, but they will not have the background knowledge they need in order to be fully involved in the others' lives. This diminishes their ability to evaluate the experience.

This was most clearly seen in our study of Photo Presence (Bentley and Metcalf 2009b). We recruited participants in social groups but also allowed them to see the public photos from other groups. The ways in which participants interacted through pictures with those that they knew was much different from the ways that they interacted through pictures with strangers. With those that they knew, participants posted photos of "inside jokes" or visuals to help explain a situation. One participant posted a picture of herself in a dress for her friend to see and approve/disapprove. Another participant posted a photo from a vacation in California for his sister in Chicago to see. She loved it and told us that "There's something satisfying about the immediacy of [knowing that] right now, in California, my brother is doing this." In contrast, interactions with unknown people were about finding pictures that were more artistic or represented shared topics of interest. Without recruiting groups of close friends and family, we would have missed out on the uses of the system that promoted sharing visual scenes in situ with close friends and family and the joy that our users got from doing this. The resulting application and feature set would have turned out much differently and it likely would not have enabled these playful experiences with close friends and family, losing much of the novelty and fun of the product.

There are a few ways to recruit participants for studies like this. The easiest, but also most expensive, is to use a professional recruiting firm. Over the years, we have found firms that are good at recruiting a broad range of participants in social groups, even when distributed geographically (in our study on communication across generation and distance (Bentley, Harboe, and Kaushik 2010; Bentley and Basapur 2012), we needed participants in South Florida who had adult children in Chicago). However, the cost of recruiting participants can often be the largest cost of executing a field study. When using a professional recruiting firm is not an option, more traditional snowball sampling can help to find participants (Goodman 1961). In snowball sampling respondents are asked about their acquaintances who might also qualify for the study. Alternatively researchers can start from their own friends of friends, although this may tend to overly bias the sample. We have used Craigslist, put up posters near college campuses and grocery stores, or even used our second-degree friend networks for some studies. In general, the goal is to find participants diverse in aspects that matter to your eventual target market. Indi Young's (2008) book *Mental Models* has a good discussion of recruiting and ways to ensure a broad sample, including useful Excel templates to use while screening participants.

Another criterion to consider while recruiting is the strength of the ties of your participants. It is tempting to want to recruit only participants with strong relationships who call each other regularly and visit in person often. Users like this will often be the most likely to be excited about new social media systems and will provide all manner of rich usage data in a trial. However, the majority of social relationships are not like this, and if the target market for a concept is a broader population, it is usually a good idea to recruit participants with a variety of communication patterns and tie strengths. For example, in the Motion Presence study, we used a group of four friends as well as three different couples in various stages of their relationships. In other studies, we have explicitly recruited based on the number and variety of communication types used between the potential participants and then screened to obtain a rich variety. As with the early-stage generative research, the more patterns that can be spotted within a wide range of participants, the more likely it is that a concept will prove to be more widely successful.

Real Context of Use

When evaluating a new mobile service or application, it is critical to see how it is used in everyday contexts. This means conducting the field trial in as natural a setting as possible and over a time period in which use becomes more natural and less exploratory. While there is much that can be learned from a quick trial in a lab or from a few days' usage, we have found that we receive the best data from multi-week field trials of use in daily life. When we put prototypes in the field we generally do so for at least three weeks. While we ask our participants to use the application or feature "enough to evaluate it," we do not ask them to use certain features or functions any number of times, on certain occasions, or for any particular reason. Indeed, we tell them that they should use it as if they are not in a study and only as they see fit. Part of our research objective is often discovering when they think it is appropriate to use the prototype (or not appropriate to use it as the case may be), and what they are interested in using it for. This ultimately will tell us if we are on to a winning idea or if we should try something else.

For example, in the Motion Presence study, we were able to see many unexpected uses of the motion information because of the real-world scenarios that our participants found themselves in throughout the trial. By seeing if their friends or close family members were at a place or moving between places, participants were able to coordinate arriving at the same place at the same time because they could see when the person who had to travel the longest had left. They were able to know that loved ones were safe because they started and stopped moving at the times that were typical. And they were able to stay aware of their friends' activities, whether it was going out on a Friday night or leaving for work very early in the morning. These cases could not be observed in a lab study or by asking someone to hypothesize how he might use such a system. It was the real use in context that allowed our participants to more fully explore the potential uses of the concept and for us to see a wide range of rich behavior that the motion data afforded. From these observations, we were able to design much more broadly for the types of use that were possible, and we decided to include motion information in concepts for calendars, dashboards, and the phone book because of these additional uses. The use of this rich presence data for so much coordinating led to many concepts for the phone book in the MotoBLUR product, where the status of contacts can be seen before placing a call or sending a text message as well as on the home screen.

Likewise in the Photo Presence study in 2007 (Bentley and Metcalf 2009b), we saw participants using the application to share images of inside jokes or just typical imagery of what they saw and experienced throughout the day. These real contexts of use allowed us to see how visual images shared in real time allowed for a whole new type of visual communication. Seeing a photo that was just taken across town or around the world and combining it with communication over the phone or through titles and comments on the media itself allowed for a participant's current environment and situation to be easily understood and appreciated by those in her life. If we had tested just in the lab or just around the home, these rich scenarios would never have arisen and our understanding of the power of instantly sharing a view of one's world would not have been included in future concepts such as MotoBLUR. In the end, photos were a large component of the Contacts 3.0 concept and eventual product. We wanted to make it easy for friends and family to experience each other's environments in as close to real time as possible. To achieve this we created a push connection of photo and status updates that are displayed in a dashboard on the device's home screen, a concept we called the "Social Dashboard" that made it to product in a widget called Happenings in MotoBLUR. By just glancing at their devices, users are immediately transported into the near-real-time images of and text updates from their contacts' lives all around the world.

In a field study of the Serendipitous Family Stories system (Bentley, Basapur, and Chowdhury 2011), the context of use was again critical. Parents left video stories for their children and grandchildren in physical places in the world. The recipients could view the videos if they traveled near the places where they were stored, and they received a notification on their phone as they were approaching a story. Because the serendipitous way that users came across stories was a key part of the experience of the system, there was no other way to test the viability of this concept than the real context of use. We wanted to know if stories would actually be found during the trial and what the process would be of finding and experiencing a story in the location where it had originally happened. In the end, we found that 83 percent of stories were discovered during the four weeks of the trial and many rich emotions poured out as our participants received the stories. They reported a large increase in communication to their family members and said that they felt closer both to their family and the history of their city because of using the system. This new experience and the feedback from users were made possible by the real context of use out in the city for a month-long time period. Some of our participants reported that they would never see some parts of the city in the same way again, or would always think of their relatives when crossing a certain bridge or passing a particular landmark after participating. One participant was able to learn about her grandmother's life as a teenager and how a college that she passed on her daily commute used to be an amusement park. Her grandmother told us that before the study her granddaughter "never knew her grand-mother was human." By discovering these stories in the places where they occurred over fifty years ago, her granddaughter was able to better imagine the life of her grandmother and the places in the city that had been most important to her as a teenager growing up in Chicago. These rich interactions with the city and personal contacts could only come from use in real situations.

Because mobile devices are meant to be used in everyday interaction in the world, it is critical to get this part right and to ensure that a new concept is tested in daily life. Part of ensuring this is determining the appropriate length of a study. This is quite often dependent on the functionality that is to be tested, the typical frequency in which users are expected to utilize the new system, and the time of year when a particular field trial is taking place. In general, a field trial should last long enough to overcome novelty effects and achieve a realistic pattern of usage. Depending on the concept, this can take anywhere from a few weeks to several months. That being said, any data that is captured, even in the first few weeks, will teach the research team quite a bit about the breadth of uses to which potential users will try to appropriate the system. We have found that steadier use usually emerges after the first two weeks of a study and that conducting studies lasting a month will be more likely to generate results that approximate sustained use. However, if researchers are more interested in initial use, shorter studies can suffice. They may be especially relevant, in fact, as research by Pinch Media (2011) has shown that mobile applications have an exponential decay with fewer than 10 percent of users continuing to use new applications after ten days. Of course, when designing compelling experiences it is important to achieve long-term frequent use, but understanding initial behavior is also quite important, especially for systems created for specific events or meant to be used in certain rare occasions such as sporting events or concerts.

In asking participants to use the new service in their daily lives, it is also vital that researchers consider the routines of use in the design of their service and field study.

New concepts should allow users to get through a day without needing to recharge their phones. To solve power issues, in small-scale studies participants can be given larger batteries such as the 2800 mAh battery from Seidio (2010) for the Motorola Droid or Google/HTC Nexus One devices. This can often be more efficient than focusing on power optimization at such an early stage in the research or product development cycle.

One thing to note when developing applications to be used in daily life is that phones are often lent to others to make calls or used in social settings where others can see the screen on the participants' phones. Consolvo and Walker (2003), at Intel Labs, make this point strongly when considering the development of systems that show personal data such as health or fitness information—when friends borrow your phone, it is likely undesirable for them to see the intimate details of your wellbeing.

Only by testing concepts in the real context of daily life can we understand how a new application or service fits into the patterns of life. By seeing the ways in which the concept is used or ignored in various situations, research and design teams can better understand the utility of the core experience they are trying to create. The data from these studies illuminates the everyday interactions that are possible only when the application is deployed in the variety of situations that one finds herself in throughout several weeks of life.

Primary Device

We also believe that in studies of mobile experiences, making the participant carry an extra device seriously detracts from the ecological validity of the research. When the final solution is envisioned as something that would exist on a primary device, having a participant carry around and use a second device will greatly impact the use of the new application or service. Use will be different on a second device that is not looked at frequently throughout the day for other purposes. Furthermore, the participant will be less likely to carry around and use the second device, especially if the only thing it does is the new experience we are studying. In sum, the participants will be less likely to engage with the prototype the way they would outside of the context of the study.

In order to maintain our one device principle we either use the participant's own device (with the application installed by the research team) or we transfer the participants' SIM cards (along with their contacts and all other information they need) to the prototype device. Transferring users' information to our own devices makes our job more difficult but also makes engaging in the study less work for the participant, and again, more closely simulates a real-life experience. By having the new experience on a device that they are using for daily phone calls, text messaging, and email checking, participants will have the opportunity to interact with the new application during these times when the phone is out and in use. Any information displayed through a widget on the home screen will be seen in these situations, thus affecting how the new application or service is used. Also, opportunities for spontaneous interaction are more likely to arise simply by having the app on the primary phone that the user has with her all the time.

Some studies that will work on almost anyone's phone include SMS-based applications. Our Music Presence application worked this way, as has one of Fogg and Allen's (2009) applications that focuses on improving health behaviors. These can be quite easy and cheap to field since there are no device costs involved.

With the growth of app stores such as the Android Marketplace and Apple's iOS App Store, it is also becoming possible to allow users to download applications directly to their own personal phones for field studies. We will address some of these issues in the final chapter, but in general this growth of easy means of distributions makes it easier than ever to issue new applications and services to larger groups of people for use in daily life on their own primary mobile devices. Platforms such as Android also make it easy to install applications on existing phones on a small scale without large distribution through an app store. In the Family Stories system, many of our users already had Android devices and we were able to easily copy the application bundle to their devices, saving the cost of purchasing a new phone for each participant. It is now becoming easier to find participants who already have a compatible mobile device, although finding an entire social group that all has the same type of device can still be difficult. We do not see the need to loan out specific devices to some participants ending anytime soon.

Field-Based Data Collection

Users in a field study are using the new application or service throughout all hours of the day and night. It is impossible for researchers to literally see and observe these interactions as they would with more traditional ethnographic field work. Over the years, we have explored a series of methods that get us closer to being there for each interaction and understanding how, why, and where certain features are used (or not). Some of these methods come from a variety of existing disciplines including anthropology, computer science, and psychology. Together, they can help us to get a better idea of how a given interaction fits into a user's environment.

Most of our studies begin with a semi-structured interview to understand current practices in the topic area in which we are interested. This could be developing an understanding of current photo-taking behaviors, phone call frequency, or social network usage. As researchers, we want to have an idea of what participants' lives were like before using the new application and to understand a bit about their social networks and expected communication patterns. These interviews tend to be quite short, around a half an hour. The data collected in this interview will help us to put into context information that we receive later on in the study and will lessen the chance for misinterpretation.

While a study is ongoing, several methods are useful for understanding particular interactions and usage patterns over time. The most basic is adding instrumentation to the application. We can log every click, save screen shots, or just log specific actions of interest such as commenting on a photo or viewing a contact list. This log data can be useful for several purposes. First, it can be used to remind participants of particular interactions in later interviews. As specific interactions are logged with the time and date, this information can often trigger a memory of value in a later interview, especially if the given interaction has been typical and unmemorable without some additional context. If location is captured with the log, this can also help participants to remember the context of use. Being able to remind a participant by saying things like, "Remember on Tuesday morning near North and Clybourn when you . . ." can greatly help in aiding recall of incidents that occurred during the study. Log data is also useful in accurately understanding frequency and times of use for the new application. Asking users, even daily, how often and for how long they used a given application generally leads to a wide range of responses that frequently differ from actual usage logs. Having precise data can help in understanding discrepancies and in building stronger models of use for later iterations of the system. Getting this log data to researchers in near-real time can also be quite useful. For the Serendipitous Family Stories study, logs were uploaded to our servers hourly if any interaction occurred during that time. First, this allows researchers to see if users stop using the system all together which could initiate phone calls to see if any technical issues might have appeared. Second, any interesting uses can lead to follow up phone calls to users or notes to add to interview protocols for mid- or end-of-study interviews. After the study, a detailed analysis of logs can give a better understanding of which functionality was used and can help in redesigning the flow and layout of the system in later versions.

In addition to automatically captured logs, hearing from users on a regular basis throughout the study allows for interactions to be explained more closely in time to their occurrence than end-of-study interviews do, since those might be conducted weeks after a particular use when the details might be long forgotten. Depending on the anticipated frequency of use, we have had participants call in to voicemail systems to discuss their interactions on a daily, weekly, or after-each-interaction basis. Longer frequencies between calls have tended to lead to less participation, as it is harder for participants to remember to call every other day or once per week. Automatic reminders on the phone could help to mitigate this problem and can be set as alarms or popups after particular interactions with links to call into the voicemail system.

In general, we have found that our best data comes right after a particular interaction when the reasons for use and details of the interaction are fresh in our users' minds. In these voicemails, we try to have our participants describe the details of use that are not captured in our logs and give enough information to correlate a particular interaction to our logs. Often this is the time and location of the interaction. We also want to know what prompted the use. It can often be as simple as seeing a notification on the device or thinking about another person because of some environmental stimulus, but this is all information that is important in helping us understand how the application fit into that particular context of use. These voicemails can then be listened to and follow-up questions can be asked at a final interview or explicitly addressed in email or phone interviews throughout the study.

We have also found it useful for participants to keep other types of logs of use. This can include taking photos at locations where they have used a part of the application or videotaping interactions with the system. This additional information can better help researchers to understand the environments of use. Is an application that relies on voice inputs or outputs being used in loud environments? Is an application that requires a lot of navigation or text entry used while carrying groceries or while standing on public transportation? These types of observations can help improve the experience so that it better fits into real contexts of use.

In some cases we have even had participants record their phone calls for us or save emails and text messages (of course with their permission and the permission of the other parties in the communication). Participants may not be completely honest in an interview with us, and how they talk about the application with others is often quite telling. Also, hearing how conversations begin or end based on information from a new experience can better illustrate how a given interaction ties into initiating social communication such as prompting a phone call or email. In the Motion Presence study, we had our participants record their phone calls with other study participants and were able to hear conversations that started with phrases such as "I saw you were on your way so . . . ." This data from phone calls is about as close as we can get to true observation in context and has proved to be quite helpful in understanding use and how experiences extend beyond the application that we have created into conversations and everyday life.

As mentioned earlier, other techniques such as experience sampling can be used to understand given contexts or how users might react in particular situations. Consolvo and Walker (2003) used this method to see what types of location information users would share with particular friends and family members when in different contexts. Throughout the day at random times, their device would alert them and ask what granularity of location they would want to share with a particular person at that time. Gaining this data in a real context throughout the day showed insights that could not have been discovered in a lab setting. ESM can also be used to get feedback from users to verify that they have an understanding about the current state of a system or are aware of the latest data that has been presented to them. Asking participants in a wellbeing study if they currently know their weight or could estimate their current step count for the day are ways to understand if the system is working and if participants are noticing and reflecting on the data that has been presented to them.

Of course, we still rely heavily on post-study interviews. These interviews not only get user opinions on various parts of the system, but seek to follow up directly on data captured throughout the study. Asking follow-up questions on data collected from a voicemail or log allows us to clarify those interactions where ambiguities were present. Final interviews allow us to gain a clearer understanding of particular interactions that the participants had with specific features of the application or service that is being tested.

Our experience is that a mix of methods for in situ data collection gives us the information we need, relatively close to the time of actual user interaction. We believe that these techniques are appropriate and essential ways of collecting more accurate data than is gained by relying solely on interviews that often occur days after particular interactions have taken place, or by just using logs of interaction that give a lot of the quantitative details of use but none of the qualitative reasons for interaction that are critical to understanding how a given concept is being used.

In the end, often a decision needs to be made on whether to continue on with a concept or to cut losses and try something else. This is frequently a hard juncture as teams may be deeply committed to the idea by this point. We try to be objective and deeply explore how often the system was used, what types of use it was appropriated for, and the general reaction of participants. Did the system really change their lives in some deeper way? Did they connect in new ways with people and places in their lives? Were they happy to use the application and sad to see the trial end?

These are all ways to get a feel for the potential future success of a system at this stage. More functional prototypes and field trials will be required to understand potential markets and user segmentation, but this stage should be about developing an understanding of whether the system is useful for people. If so, moving on to larger beta trials and commercial launches is often the next step and will be the focus of the next chapter.

Discount Methods

While we have found that running a full multi-week field study can give us the best data on how a given concept will be used, depending on the application there are ways to shorten the process and still obtain useful results. This can involve running a trial at particular events or recruiting participants who are about to embark on an activity of interest (e.g., planning a weekend trip, meeting friends in person, choosing a movie to watch, etc.).

For the TuVista sports video system (Bentley and Groble 2009), we conducted several trials at live sporting events. These were often single-game events and participants were recruited from those attending the game on a given night. While there was certainly a novelty effect, we could see the particular ways that on-demand instant replays could be used throughout a game and could directly observe users and their (often quite emotional) reactions over time. When studying users in a given interaction, it is important to observe as much as possible how the system is being used, and therefore during the TuVista trials, we had researchers stationed near participants observing their interactions. After the event, participants were interviewed, using cues from these observations. One particularly moving scene was a group of male fans in their thirties holding the phone high above their heads after the game while replaying the winning goal. As it played they danced arm in arm while loudly singing the team's victory song and quite naturally holding the phone up high for everyone to see the reason for their celebration. Gaining insight from these very emotional interactions can often happen rapidly where direct observation is possible. These observations meet with Grimes and Harper's (2008) ideal of "designing the experience of [an] application such that the technology [itself] is 'lost' in the moment," and thus is a natural support to the celebration.

Nakhimovsky et al. (2010) at Google conducted similar trials when testing the Google Navigation product. They recruited Google employees to use the system on their daily or weekend drives and provide feedback based on their use in these already occurring situations. For systems that are meant to be used in particular situations, observing users in these situations can help get initial data on how the new experience fits into these activities.

Additionally, our students have found friends who are about to plan a trip or go grocery shopping in order to perform quick one-time trials of their systems during the semester. While not as informative as a multi-week study, they were able to learn quite a lot about the use of their systems in real contexts. Because some activities do not occur every day, these quick one-time trials can be a useful way to quickly observe use in real-world scenarios.

These types of evaluations can form a good start to understanding use in practice. However, longer studies that attempt to overcome novelty effects are still necessary when you are concerned with long-term use of a new service concept and how it is used after the initial novelty wears off. They can give development teams the data they need in order to plan new versions of the concept and to determine which aspects of a concept are worth pursuing further. We recommend many iterations of this testing in order to better understand how new concepts can evolve to meet the many uses and contexts of use that users can imagine for a new application or service.

9 Distributing Mobile Applications: Putting It All Together

The end goal of most mobile development efforts is to create something that many people all over the world can use. Taking an application from a small field study to a larger distribution often requires quite a bit of effort, both in hardening the system itself and in the actual distribution of the application. There are many factors to consider when distributing an application. Instrumenting the system to collect the data that you will need to analyze users' interactions and improve your offering is one key issue, but along with this comes ethical questions of collecting data from large numbers of users as well as scalability issues. In addition, there is the problem of interpreting all of the data that is collected. Is initial usage typical of the demographics of the wider population? How might use change as more people sign up? Knowing the answers to these questions can make the difference between a successful offering that scales and responds to user demand and a system that fails to attract new users or cannot keep up with their usage.

Releasing mobile applications and services used to be a difficult process. Before the rise of the current carrier-agnostic application stores, application creators usually faced two main paths for distribution: they could post applications on their own websites and attempt to draw users to download the apps to their devices (often by typing in long URLs), or they could partner with an operator or phone manufacturer to get the application preloaded on certain cell phones or put in a carrier application store (usually branded as "Games and More," or similar monikers). This left app creators and potential customers often unable to find each other. Working with carriers and phone manufacturers to preload applications often took over a year due to the release cycle of mobile phones and the various legal agreements that had to be made.

Yahoo! and Google were two companies that quickly explored alternate ways to launch mobile applications. ZoneTag (Naaman, Nair, and Kaplun 2008) was a project created at Yahoo! Research Berkeley to tag and upload photos to Flickr that was first released as a public beta for Nokia Series 60 devices, and later in collaboration with

Motorola for Motorola devices that supported J2ME applications such as the RAZR 2 and SLVR. To download the application for any platform, users visited a website hosted by Yahoo! and entered their mobile phone number. A link to download the application would be sent to the device and users could follow the installation process for their particular platform. The photo sharing service Radar (Garau et al. 2006) worked in a similar way. Some phones made this easier than others, but for many users, this was the first mobile application they had ever installed and there were many points at which they could get lost along the way. Google followed a similar path with their release of J2ME applications for Google Maps and Gmail (Google Gmail Mobile 2011). Links to download the apps were placed in the mobile web versions of these products, and the Google website allowed users to send the URL for the mobile app to their device for download. This process helped Google to quickly grow the use of their mobile applications by users that they already had on the mobile web. Still, discovery of the applications by end users was difficult as there was no central place to visit to learn about all applications that were available for a given device.

Today, the landscape is quite different and major application markets exist for almost every mobile platform. As of this writing, the Apple iOS App Store and Google Marketplace for Android devices were the largest stores and gave users of iPhones or Android devices easy access to browse catalogs of hundreds of thousands of applications. Both markets allow for user ratings and comments for applications as well as the ability for users to find the most popular applications in a given category. For developers, getting an application hosted by one of the major markets is an easy procedure. However, there are still some issues that should be considered before posting an application for access by the world.

This chapter will cover some of the most common issues that are faced when launching a new mobile application or service, including strategies for prelaunch betas, scalability, instrumentation, ethics of large-scale research, mobile business models, and upgrades. These are all important topics to consider when moving beyond small trials toward wide deployment of a new concept.

Beta Releases

Before an application is released to a potential audience of millions of users through an app store, it is usually beneficial to run a trial in some smaller scale. Often, this is larger than a field trial of a dozen or so users but smaller than the hundreds of thousands or millions of users your application might see in a full public release. A beta release is often intended to identify any final major bugs in the system with the help of a reasonable number of new users or to collect usage data on a late version of a system to help plan for scaling the final solution.

The different mobile platforms allow for limited releases in different ways. The iPhone platform allows for ad hoc distributions for up to one hundred devices. This type of distribution requires that all of the devices are known ahead of time and that a provisioning profile can be obtained for the devices before the trial is conducted. Still, this can often be the best opportunity for testing a new product with a small set of users.

In the TuVista project in Motorola Labs, Frank and Mike Groble (2009) used the ad hoc distribution method to test the TuVista application at UCLA volleyball games and at the Paralympic World Cup in Manchester, England, before its larger commercial release with the Denver Broncos. In addition to simply being initial trials with potential customers, these early trials were also designed to obtain feedback from sports fans in order to understand how the system could be improved for a final commercial release. For these trials, we supported up to one hundred users in order to get some larger-scale data from fans at these events and build more accurate usage models.

Due to the ad hoc deployment limitations of the iPhone SDK (iTunes Store Terms and Conditions 2011), for these trials we had to recruit participants in advance and meet with them before the game started to install the application on their devices through a cable (Distribute Your App 2011). Once the game was underway, our participants were able to use the application in an unrestricted manner and we were able to learn quite a bit about user behavior and traffic models so that we could better plan for scalability when it came to a full release. Through surveys, observations, and interviews associated with these trials, we were also able to make key improvements to the user interface to help people get to the content they wanted as easily as possible, especially after thousands of clips were posted to the system for over a hundred events in the Paralympic World Cup. New views were added so that users could easily find the newest as well as the most popular content across all events, which was quite important as the number of simultaneous and overall events increased.

The Android operating system makes public betas a bit simpler (Google Apps Marketplace Terms of Service 2011). The application bundle, or APK file, can be hosted on any server. As long as the phones are allowed to download applications that are not from the Google Marketplace (as of this writing, some operators were imposing this restriction while others were lifting it), users can just click on a link to the APK file on the web and install it on their phones. If necessary, this file can be kept on a password-protected or Internet Protocol (IP)–filtered site to prevent wide distribution beyond the intended beta trial. This is how we released Serendipitous Family Stories for our twenty-user evaluation (Bentley, Basapur, and Chowdhury 2011) and for an internal prerelease trial at Motorola. APK files can also be directly emailed to participants and installed right from the Android Gmail application on their phones.

Other researchers have moved beyond closed, private beta trials to larger open trials where their systems are available on the Apple iOS App Store or Google Marketplace. In the Phi Square system designed by researchers at the Mobile Life Centre in Stockholm (Büttner et al. 2010), visitors to a website can create physical tags (made of a Quick Response (QR) code) for places in the world. Using a mobile phone app available in the Google Marketplace, users can scan the tag and with one click check in to the venue represented by this tag in FourSquare, a mobile location check-in service. This website and mobile application were released publicly for anyone to use and in the first month of release six hundred physical tags were created. When appropriate attention is placed on the scalability of a system, such public betas can help researchers obtain a large sample of meaningful usage statistics and learn about the types of places where such systems are used.

Andrew Campbell and students at Dartmouth (Miluzzo et al. 2007) have released their CenseMe application through the Apple iOS App Store. With this application, the current context of a user can be determined and shared to Facebook. Through this deployment, Campbell and his students were able to test their system in many different scenarios that would not be possible to cover in a small-scale trial.

Donald McMillan (2010) and colleagues from the University of Glasgow in the UK have field-tested several popular applications in open trials in order to perform large-scale research. Their Hungry Yoshi game has been downloaded over two hundred thousand times and their World Cupinion system was heavily used during the World Cup to vote for teams that users thought would win in various matches. Through these large deployments, they have been able to create tools such as SGVis (Morrison and Chalmers 2011) that allow for analyzing usage in real time across multiple user categories or by following one particular user over time. Tools like these will be instrumental in understanding larger deployments as more applications begin to reach ever-increasing audiences.

Scalability

While many mobile experiences can work in a client-only or peer-to-peer mode, most will require some sort of backend server in order to support their interactions or data collection. When putting an application into an app store, preparing for a large number of users is often one of the biggest concerns. Today's cloud computing systems such as Amazon EC2, Microsoft's Windows Azure, or Google App Engine can help mitigate this risk with the proper design. These services allow for easy porting of server code and data to classes of servers with different computation and memory support. They also provide load balancing services to divide requests across multiple servers if usage gets high enough.

Design for scalability needs to be an early part of the system design. Databases must be able to be run across multiple servers and users must be able to be provided with consistent data based on load balancing policies. Luckily, most of this work is the same process that would need to be followed for any new website that is about to be launched and many resources exist to help developers better plan for scalability. Barr (2010) and Henderson (2006) provide additional details for architecting the web service components of a system to better scale to increasing demand.

To help with scalability, content can be cached on devices for later retrieval. When content is cached, it does not need to be re-downloaded each time a user requests it. Also, because mobile devices are often in areas of minimal connectivity, as mentioned in chapter 5, it is often useful to have data already available on the device. Larger downloads can also be scheduled when the server does not have peak demand, such as in the middle of the night. By using conditional GET requests from the mobile phone, re-downloads will only occur when the content has changed. This is a good strategy when interacting with a system that is expected to be busy, such as with live-event content, and when the content is not likely to change. However, if users typically download only a small portion of the content, or if it cannot be predicted ahead of time which content they will likely explore, pre-caching content can waste a good deal of bandwidth and server resources. It is important to consider these aspects when choosing a content delivery strategy. Other options include using content distribution networks such as Amazon Simple Storage Service (Amazon S3), which handles the delivery and global caching of popular content files.

Instrumentation

When applications are released in the wild, research and product teams often want to know how the system is being used. This generally involves instrumenting the application with hooks to record desired user interactions and share them with a server. In most cases, combinations of mobile application logs and server logs can be combined to form a complete picture of a user's interaction with the service.

Instrumentation can help to discover features that are the most popular, discover where paths through the application are not optimal for tasks that users are performing most frequently, and identify areas that can be improved or might need better prompts or labels. Usage data can also help engineering teams to plan for future scalability or to pre-cache data that is likely to be used in order to provide for faster screen loading times.

Often, transitioning to different screens of an application will result in unique requests being made to a backend server. In these cases, traditional web server logs can capture a lot of the usage of many mobile systems. These requests can be analyzed to understand feature usage as well as patterns of interaction over time. Combining server logs with download counts from app stores can help research and product teams to understand the percentage of users who open applications once downloaded and their usage patterns over times of day as well as over longer periods of time. This data can help in understanding the scalability needs of a system and the patterns of demand that are created.

However, the server logs show only one side of the interactions that occur with the system. Often, it is important to know about the use of the client application, especially when many user interactions do not directly result in a server request. In these cases, mobile applications need to be instrumented to collect data about their use. This instrumentation captures user interactions, screen loads, button presses, etc., and periodically sends this information to a server for analysis by research and product teams.

In the Motion Presence system described in chapter 4, Frank and Crysta Metcalf (2007) added logging capabilities to keep track of when users opened and closed the augmented phone book application as well as when they clicked on a particular contact. Any instances of phone calls and text messages initiated from the application were also logged. This usage data was stored in a flat file on the SD card and retrieved before the final interview at the end of the study, but also could have easily been uploaded to a server if used in a larger distribution. From this data, we were able to learn about the times of day when users checked on the status of their friends and family, seeing spikes during lunchtime and after work. We could also see instances of interaction that were not reported in the voicemail diaries and follow up on those. This helped to give us a better understanding of the use of the system and the value that it was providing for our users.

Similarly, in the Serendipitous Family Stories system (Bentley, Basapur, and Chowdhury 2011) we logged each instance of opening or closing the application, times when a particular story was viewed, and instances when notifications of nearby stories were viewed or hidden. We also kept track of video plays and instances of communication initiated from the application. The data was stored locally in a flat file and then periodically uploaded to the server throughout the day. This data helped us understand how users uncovered new stories as they went about their lives, as well as how already discovered stories were replayed and shared with others at later times. In addition to learning how stories were uncovered throughout the month-long study, we also observed increased patterns of story finding on weekends when participants could devote time to wandering around the city in search of nearby stories.

Instrumentation can be a critical part of understanding use in systems that are deployed to large numbers of people. When it is no longer feasible to meet with users individually, logs of interaction can help show how applications are being used, at least. But when looking for the "whys" of interaction, methods like those introduced in chapter 8 are still beneficial in understanding users' motivations for using the system in various ways. Ames and Naaman (2007) from Yahoo! demonstrated this nicely in their study of the ZoneTag system. Usage data from months of deployment was combined with interviews of a small sample of current users in order to discover the motivations users had for tagging content. This led to deeper insights about the use of the system and identified four main motivations for tagging content. With this data, future systems could better support users' approaching the system with any of these motivations. Only a combination of qualitative and quantitative methods could allow Ames and Naaman to come to these conclusions.

There are several other important considerations to keep in mind when creating log data. It is helpful to think of your research questions or product goals when determining which data should be captured. It is also important to take into account the privacy of your users. If potentially sensitive data is being collected (such as location, user IDs, etc.), special care should be taken to properly anonymize, hash, or encrypt the data that is sent over the network. In general, though, time stamps are critical on all user interactions to help understand flow and the context of particular actions. Requiring log uploads to be cryptographically signed like other sensitive API requests helps prevent fraudulent log upload data.

Furthermore, it is often desirable to analyze usage information according to various demographics. Research from Telefonica (Church and Cherubini 2010) has shown that for new systems that see early adoption among concentrated demographics, simply looking at statistics across all early adopters will not give a realistic view of how more typical demographics will use the system. Properly weighting responses according to target demographics will give a better indication of the system's use once it has crossed the chasm (Moore 1991) between the early adopters and the mass market.

We often write small programs to analyze log data, pulling out statistics on per-user and per-content item use. We can then perform further statistical analysis on the results or plot them to see patterns that emerge or how use evolves over time. This data can help us understand long-term use and interactions across users that would not be as easy to obtain through qualitative means. Other systems, such as Morrison and Chalmer's (2011) SGVis, can help to sort through the amounts of data that are coming in from larger deployments where writing manual scripts to parse through huge log files might become unwieldy. These tools also allow researchers to look into specific data for individual (anonymous) users and categorize use so that they can better understand how a system is being used in the world.

Ethics of Large-Scale Research

Small-scale research generally means that researchers get to meet with each of the participants and explain the purpose, benefits, risks, and data collection procedures of their work. Through informed consent procedures, participants in the study are given a chance to read about the details of the research and data collection and discuss them with the research team before engaging in the study and while the data collection is occurring. These policies of informed consent have been derived from years of best practices in social science research and have been created to protect the privacy and experiences of participants in research studies. Any potential risks the user might face are clearly stated. Often, these consent forms and research protocols must pass stringent Institutional Review Boards (IRBs) before the research can be conducted in order to ensure the privacy and wellbeing of the participants.

However, large-scale research by definition means that the creators of the system or scientists running a study are not able to meet with all participants. A description on a website or End User License Agreement (EULA) is often all a user has to read when choosing to install a new mobile application. On the Android platform, they can also see which platform services the application has access to (e.g., Location, Phone Book Read/Write, Data Services, etc.) but have no way of directly knowing what data is shared over the network and how it is stored.

Studies of EULAs have shown that the majority of users do not read or understand them. In a study of 222 users, Nathanial Good and colleagues (2007) from UC Berkeley and the University of Minnesota found that providing a summary of EULA conditions in addition to showing the license significantly decreased the number of installations of particular applications. This study highlights the issues that teams face when launching an application. If users are fully informed, they are less likely to use a new service. However, not fully informing them leads to misunderstandings about the system they are about to engage with and counters years of ethical research practices.

How can users be fully informed of the terms of new applications if they do not read EULAs or do not fully understand them if they do? Several research projects have shown innovative ways to present terms of use to users as they are installing an application. Moving beyond the Google Marketplace's limited view of permissions and data that can be accessed (Manifest.permission 2010), Good et al. (2007) provided a summary of how collected data is used on a separate screen for the user. For example, users could be informed that particular data would be uploaded and stored with their user account on a server, shared with advertisers, or used anonymously to improve recommendations. In another study, Kay and Terry (2010) created pictograms of data being shared—for example, showing a keyboard with an arrow to a server. These types of disclosures can help users to better understand the conditions of use for the service with which they are about to engage. However, as Good found, declarations like these can significantly reduce the number of users who install and use a particular application. We hope that one day disclosures like this will become commonplace so that all users are not only fully informed about the terms of an application and its corresponding service before using it, but that disclosures do not deter them from use.

Related to ethical issues is the problem of the reliability of data provided by users. If a system asks for demographic data and the research team or marketing team later reports on findings using this data, can it be trusted? Research from Telefonica Labs (Church and Cherubini 2011) has found that many users provide incorrect data for fields such as email, age, or occupation. Often, these are required fields to sign up for a new service and users have no incentives to enter accurate values (and are often in a hurry to get to the service itself). If findings are tied to these self-reports, incorrect statements might be made about the use of the system in the wild.

Revenue

As a new mobile application nears commercial release, the business model that is going to be used becomes an important consideration in the final design. There are currently several models to choose from that have been successful in the various application marketplaces. Apps can be free, ad supported, downloaded for a set price, contain in-application purchases, or be tied in to a paid account for a service that is established on the web. Each of these models has benefits and shortcomings, sometimes depending on the particular platform on which the service is launched.

According to distimo.com, as of November 2010, 29 percent of iPhone applications and 60 percent of Android applications were free to download. This included many popular applications such as Facebook and FourSquare that were also completely free of advertising. However, many of these free applications did include some form of advertising from providers such as Apple's iAd or Google's AdMob platform. These services allow application developers to include advertisements on various screens either as a banner or as a full-screen interstitial ad. Ads from both providers are dynamically priced, providing varying revenue to application providers based on the cost for each ad that is displayed. Even with small revenues per 1,000 views or per click, these systems can provide significant amounts of revenue when large numbers of people are using an application or if their use lasts for long periods of time (e.g., in a game with a new banner ad across the bottom of the screen, a new ad can be displayed each minute while sessions tend to last for tens of minutes each).

Paid applications provide another path to the market for application creators. As of November 2010, the average price of paid applications was $4.31 for the iPhone and $3.23 for Android according to distimo.com. Paid applications tend to have significantly fewer downloads and several strategies have been used to entice users to pay. Free versions of games where only the first level is available or lite versions of applications that lack key functionality have been employed to introduce users to a service before they buy a full application. As an example, the Android application PDA Net (2010) allows users to connect their computers to the Internet connection of their phones. The free version only allows for unencrypted HTTP traffic, but the paid version allows users to browse sites that use HTTPS (Hypertext Transfer Protocol Secure) such as Gmail or other sites that encrypt their authentication pages. In this way, users have a risk-free way of determining if the app will work with their devices and testing the typical data transfer speeds that can be provided through phone tethering before deciding to purchase the fully functional application. Reviews and comments in the app store marketplaces also help users decide on the value of an application before choosing to pay for it.

In-application purchases provide a third means of obtaining revenue from users. Applications such as MLB At Bat (2010) for the American professional baseball league allow users to purchase live video streams of a current game for $0.99 each. Systems like this allow for additional revenue after the application has been downloaded. If well-presented and linked into simple payment systems (such as the one-click purchasing in the Apple iOS App Store), this strategy can lead to an easy way for users to purchase premium content without downloading new applications for each purchase.

Finally, some mobile applications are distributed for free but work with a web-based service that users are already paying for. Applications such as Pandora (2011), Spotify (2011), Netflix (2011), and Skype (2011) tie into accounts that users already have via their website login credentials and provide content or communication that draws on their existing service plans without additional payment on the mobile device. In this way, the mobile application can extend the reach of the website to a new screen in the user's life. It can also take advantage of functionality not possible in a mobile web application such as background processes, camera, and sensor access.

The choice of business model for a particular application is often complex and depends on the wider ecosystem in which the new application or service is used. Taking into account pros and cons of each distribution method for a given application can help providers to create an offering that works best for their users as well as their own business needs. Anticipated session length, usage frequency, and larger ecosystem interfaces all form key components in determining the price and ad strategy for a given application or service.

Upgrades

Once an application is launched, it often needs to be maintained and updated. As new releases are made, users often face the choice of upgrading or not. In most cases, old versions need to be supported indefinitely as many users choose not to upgrade. Often, adding advertisements to an existing ad-free application can cause many users to choose to skip an upgrade. The developers of the mobile game Galaxy Impact (Wang 2009) advise application publishers to include ads from the start if that is the desired business model, as they saw a steep decline in upgrades and new downloads after turning on advertisements for their game via an update.

Accordingly, older API versions often need to be supported indefinitely since it is likely that some number of users will not update the application. This places additional burdens on server-side infrastructure in mapping older API requests into current database schema and representations that evolve over the course of the product's lifecycle. Making initial APIs generic enough to support a variety of web-based content delivery and display can help reduce the need of refactoring in the future, but places constraints on the interactivity of the resulting application and use of platform-specific UI controls.

Larger concerns often arise when considering the end state for a given service. When it is time to end a large-scale public research project or give up on a new product that is not performing well, the service often needs to be shut down. This can mean providing simple ways for users to download all of their data or port it to another service. It can also mean keeping the service running on inexpensive hardware for existing users. Even though there were no clients for modern smartphone operating systems and the research institution it was developed in no longer existed, the ZoneTag system from Yahoo! Research Berkeley (Naaman, Nair, and Kaplun 2008) ran until late

2010. A system developed by Nokia and UC Berkeley (The Mobile Millennium Project 2011; Amin et al. 2008) to show travel times on San Francisco Bay–area roads is also still in operation for users who had accounts at the time the service was closed to new users.

In these cases, it was relatively inexpensive to continue the service, but in many cases this may not be practical. Considering the end-state of a new service is often quite important and the commercial state of a new system is something that should be mentioned to users upon starting to use a new service. Is the system a research prototype that likely will be taken down? Is it a first release of a product that may or may not succeed? Or is it an established business? These factors could influence how much effort users put into creating content in a system. Good data portability policies can ensure that they will be able to move their data elsewhere should the service fail, and these policies should be clearly stated for new users.

***

Fortunately, app stores have made it possible for anyone who can create an application to release it to an audience of millions of potential users. With the right planning, this can be a smooth process and the new experience that is created can be available to people worldwide.

Mobile phone development and commercialization has changed quite a bit in the past five years, but the desire for people to be connected to each other and to their information wherever they are has only grown. The next five years should bring about just as much change as cell phones, tablets, and televisions all converge on the same operating systems and application distribution models. This convergence of devices will bring about new types of experiences with our friends, our content, and our cities that we cannot dream of today. With the methods outlined in this book, we hope that researchers and product development teams can learn from observing others and create the next round of compelling mobile experiences that strengthen social relationships and connect people to each other and their content in new and exciting ways while more naturally fitting into their daily lives.

10 Conclusion

Creating mobile experiences that allow for new forms of interaction with the world requires more than just design skill. It requires a deep understanding of people and of the technologies that make up the mobile ecosystem. It requires building and trying and tearing down and rebuilding. Developing a new service that will fit into daily life and enhance the lives of its users is not a simple, linear, repeatable process. But we hope that the approaches and examples from this book have given you a toolbox of methods from which to choose when working on your own projects.

Mobile applications and services have evolved rapidly over the past decade, and the next decade will bring more changes as the lines between mobile device, tablet, and laptop blur even further. However, one thing has remained constant throughout this time of great change, and that is the way that the mobile device is situated in our daily lives, able to draw upon rich context and better connect us to each other and to the world we live in. Taking advantage of this unique functionality of the mobile device allows for experiences that make life more fun, serendipitous, or simply more efficient. Mobile apps and services can help us to be better people, to maintain or improve our health, to be aware of happenings in the lives of our friends and family, and to be more aware of the world around us.

We believe that the most impactful new mobile applications and services will take advantage of the unique functionality that the mobile device brings and the new contexts in which it can be used. Creating a compelling mobile application is not as simple as just porting a website to a smaller screen format and touch interaction style. Mobile applications should be built to use the unique context and environmental sensing that the phone provides. Mobile systems should tie in data from multiple sensors as necessary and take advantage of the connections between people, their close friends and family, and their information. Systems such as MotoBLUR or StoryPlace. me work in people's lives because they tie in to the ways in which people use mobile services and how they want to connect with others. For friends and close family, a feed of updates while on the go can create opportunities for spontaneous meet-ups or just a general awareness and increased connection with others. Mobile video tied to a particular place in the world helps to contextualize that place in the lives of close friends and family members, forever changing a person's future interactions with both that place and the creator of the story. Concepts like these have the greatest power to transform the daily lives of their users.

Almost none of the concepts in this book came from blind invention. They were motivated by data from real-world studies of people and their interactions with each other, their technology, and their media. Using real-world behavior as a starting point for design inspiration ensures that ideas are meeting real needs that people have and can fit into their patterns of life. As teams at Motorola and teams in our class have demonstrated, this dramatically increases the likelihood of concepts proving to be successful.

By building initial prototypes quickly, effort is not wasted in developing full systems only to have them turn out to be unsuccessful with users or technically unfeasible. Rapid prototypes allow for a quick and simple test of the core concept of a new system and help validate it in the daily lives of a diverse group of real prospective users. Often, these systems can be built in a matter of days or weeks and can massively de-risk a new concept.

No matter where one starts in the cycle, returning regularly to users to understand how a concept is adopted into their lives is the most critical step. Without this usage information, concepts cannot be properly validated and teams take a greater risk in launching them either as products or public research systems. Building compelling mobile experiences requires a constant cycle of iteration of designing, building, and testing a concept. This is a process that continues indefinitely with many systems currently on the market, leading to a cycle of improvement (or a perpetual beta depending on how you look at it).

Designing a mobile experience is not a mammoth task to be completed all in one go. Design should be iterative and follow from the results of field and usability studies. Each round of design should aim to create and utilize appropriate conceptual models for the users and navigation should flow in a way that is clear. Prompts and icons should be designed to guide users through the application and interaction should be repeatedly evaluated in real-world contexts.

The usability of mobile systems is an important aspect of design. Because of the limited screen size and interaction methods available on most mobile devices, usability is even more important than on the web where many navigation strategies can be supported simultaneously. By creating paper prototypes early in the design process and testing these with users, development time will be better spent and less revision will be needed once the design starts moving to code. Lab usability studies should be combined with real-world evaluations in order to best understand how a concept works for users in actual contexts of use.

The specific unique technologies of mobile devices and their implications for real-world use also need to be well understood in these prototypes and resulting system designs. Multiple means of determining location are available and all are well suited to specific purposes. Mobile networks may promise connectivity "everywhere," but in practice users travel in and out of connectivity throughout the day and see a wide range of network speeds depending on the network that they are currently connected to and the signal quality at any given time. Because of this, care must be taken to cache appropriate data and provide meaningful experiences when no network is available or when throughput slows to a crawl.

The ultimate test of a new mobile experience is how people use it in their daily lives. Multi-week (or month) field evaluations allow a research or design team to understand the ways in which their concept is used in real settings and how the design can be improved to better meet users' needs. Often, the uses that are observed will be surprising and can sometimes lead to an appreciation of a concept in an entirely new context or to creating new and even better concepts. More commonly, testing in real-world settings allows for a continual improvement of the concept and a focus on the use cases that are core to the experience.

Releasing an application to the public, either as part of a large public beta or as a product, requires additional work. Instrumenting a system to collect meaningful usage data, ensuring that any service components will properly scale to many users, and managing upgrade policies and out-of-date applications are all issues that teams face. However, getting a new application or service out into the world is often the greatest goal for a research or product team. Reaching thousands or millions of users and affecting the way they live their lives and interact with the people and places that are most important to them can be the greatest reward for traveling down this convoluted path of building a mobile experience.

Acknowledgments

Frank Bentley

I would like to first thank a long group of advisors and research collaborators at MIT for introducing me to the field of Human Computer Interaction and the idea that the interactions between people, their devices, and information was and is a fascinating area to study. In an undergraduate project at the MIT Media Lab with Heidi Gitelman, I first saw the methods of early prototype evaluation and from that point on knew that this was the topic that I wanted to explore. Howie Shrobe and Trevor Darrell continued my interest in this area, but it was Konrad Tollmar who introduced me to the rich literature from the CHI and CSCW communities. Together we explored interesting concepts around mediated communication.

My colleagues at Motorola have been instrumental to the success of much of the work presented in this book and the development of this research process over the past ten years. Much of the earlier work in this book was in collaboration with Crysta Metcalf, Motorola's only anthropologist. Together we've tried many methods and she's been instrumental in my understanding of qualitative methods and grounded theory. Our managers and directors over the years have been supportive of our experimentation and working outside traditional processes. Larry Marturano and Ken Douros have both helped shape our work and many long conversations have challenged me to question existing processes and develop new methods to answer our research questions and invent new applications and services. The designers at Motorola's Consumer eXperience Design over the years have shown me the importance of good design and of working through a concept before jumping to screen design. JoEllen Kames, Lauren Schwendimann, Rhiannon Sterling Zivin, and Rafiq Ahmed have all helped to expand my awareness of mobile design, and they are all-around great people to work with.

Colleagues in the CHI community have helped me push beyond existing methods and theories and have led to great collaborations. Mor Naaman, while at Yahoo!

Research Berkeley, helped me launch Motorola's first ever public beta with the J2ME version of ZoneTag. And we've been talking about social media systems ever since. Along with Mor, Ayman Shamma and I have had many conversations on the topics of human-centered research and application instrumentation, many of which have shaped the work in this book. Henriette Cramer and the group at the Mobile Life Centre in Stockholm have challenged my thinking about field evaluations, especially large-scale deployments. Much of the work in chapter 9 has come from our collaborations on a "Research in the Large" workshop and journal special issue.

Finally, I'd like to thank all of the students that Ed and I have taught over the years. They have all helped to shape my thinking of what is possible and continually remind me that a twelve-week semester really is a long time and much can be accomplished in such a short time period.

Ed Barrett

My thanks to our colleagues in the Program in Writing and Humanistic Studies and the Comparative Media Studies Program at MIT, and to our students for their inspirational creativity, intelligence, and willingness to take on remarkably difficult challenges.

Author Bios

Frank Bentley is a Principal Staff Research Scientist in the Motorola Mobility Applied Research Center with ten years of experience in the mobile industry. Frank's work investigates ways in which mobile devices can help strengthen strong-tie social relationships. He takes projects from early conceptual studies through to prototyping, field evaluation, and product as a routine process. Frank also co-teaches the MIT course 21w.789 (Communicating with Mobile Technology) with Ed.

Edward Barrett is Senior Lecturer in the Program in Writing and Humanistic Studies at MIT and affiliated teaching faculty in Comparative Media Studies, MIT.

References

About Apple Push Notification Service. 2010. Accessed October 5. http://support.apple.com/kb/HT3576.

Ahern, Shane, Marc Davis, Simon King, Mor Naaman, and Rahul Nair. 2006. Reliable, User-Contributed GSM Cell-Tower Positioning Using Context-Aware Photos. In *Adjunct Proceedings of the Eighth International Conference on Ubiquitous Computing.* UbiComp 2006. Irvine, California.

Ames, M., and M. Naaman. 2007. Why We Tag: Motivations for Annotation in Mobile and Online Media. In *Proceedings of the SIGCHI Conference on Human Factors in Computing Systems.* San Jose, California. April 28–May 3. CHI '07. ACM, New York, NY, 971–980.

Amin, S., et al. 2008. Mobile Century—Using GPS Mobile Phones as Traffic Sensors: A Field Experiment. *15th World Congress on Intelligent Transportation Systems.* New York, NY. November.

Amir, Mohamed. 2010. Energy-Aware Location Provider for the Android Platform. Master's thesis, University of Queensland.

Android Cloud to Device Messaging Framework. 2011. Accessed January 12, 2012. http://code.google.com/android/c2dm/.

Angrosino, Michael V. 2002. *Doing Cultural Anthropology: Projects for Ethnographic Data Collection.* Prospect Heights, IL: Waveland Press.

App Inventor for Android. 2011. Accessed February 24. http://appinventor.googlelabs.com/about/.

Audioscrobbler, A. P. I. 2011. Accessed February 27. http://www.audioscrobbler.net/.

Balagtas-Fernandez, F., and H. Hussmann. 2009. A Methodology and Framework to Simplify Usability Analysis of Mobile Applications. In *Proceedings of Automated Software Engineering '09.*

Ballagas, R. A., S. G. Kratz, J. Borchers, E. Yu, S. P. Walz, C. O. Fuhr, L. Hovestadt, and M. Tann. 2007. REXplorer: A Mobile, Pervasive Spell-Casting Game for Tourists. In *CHI '07 Extended Abstracts on Human Factors in Computing Systems.* San Jose, CA. April 28–May 3. CHI '07. ACM, New York, NY, 1929–1934.

Bambuser. 2011. http://www.bambuser.com.

Barr, Jeff. 2010. *Host Your Web Site in the Cloud: Amazon Web Services Made Easy: Amazon EC2 Made Easy*. San Francisco, CA: SitePoint.

Barrett, Edward, and James Paradis. 1988. Teaching Writing in an On-Line Classroom. *Harvard Educational Review* 58 (2):154–172.

Baum, L. Frank. 1900. *The Wonderful Wizard of Oz*. Chicago, IL: George M. Hill Company.

Bentley, Frank, and Edward Barrett. 2011. MIT class 21w.789. http://web.mit.edu/21w.789/www/ Spring 2006, 2007, 2008, 2010, 2011, 2012.

Bentley, Frank, and Santosh Basapur. 2012. StoryPlace.me: The Path from Studying Elder Communication to a Public Location-Based Video Service. CHI 2012 Extended Abstracts (Case Study).

Bentley, Frank, Santosh Basapur, and Sujoy Kumar Chowdhury. 2011. Reminiscing Through Location-Based Asynchronous Video Communication. CHI '11 Workshop on Bridging Practices, Theories, and Technologies to Support Reminiscence. May.

Bentley, Frank R., and M. Groble. 2009. TuVista: Meeting the Multimedia Needs of Mobile Sports Fans. In *Proceedings of the Seventeen ACM international Conference on Multimedia*. Beijing, China. October 19–24. MM '09. ACM, New York, NY, 471–480.

Bentley, Frank, Gunnar Harboe, and Pallavi Kaushik. 2010. Involving Seniors in Ethnographic-style Work: Initial Findings from a Study on Long-Distance Communication. CHI '10 Workshop on Senior-Friendly Technologies: Interaction Design for the Elderly. April.

Bentley, Frank, Gunnar Harboe, Crysta Metcalf, Vivek Thakkar, and Guy Romano. 2008. Multimedia Device for Providing Access to Media Content. U.S. Patent Application 2080060014.

Bentley, Frank. JoEllen Kames, Rafiq Ahmed, Lauren Schwendimann, and Rhiannon Sterling Zivin. 2010. Contacts 3.0: Bringing Together Research and Design Teams to Reinvent the Phonebook. In *Proceedings of the 28th of the International Conference Extended Abstracts on Human Factors in Computing Systems*. Atlanta, Georgia. April 10–15. CHI EA '10. ACM, New York, NY, 4677–4690.

Bentley, Frank R., and Crysta J. Metcalf. 2007. Sharing Motion Information with Close Family and Friends. In *Proceedings of the SIGCHI Conference on Human Factors in Computing Systems*. San Jose, California. April 28–May 3. CHI '07. ACM, New York, NY, 1361–1370.

Bentley, Frank R., and Crysta J. Metcalf. 2008. Location and Activity Sharing in Everyday Mobile Communication. In *CHI '08 Extended Abstracts on Human Factors in Computing Systems*. Florence, Italy. April 5–10. CHI '08. ACM, New York, NY, 2453–2462.

Bentley, Frank, and Crysta Metcalf. 2009a. Rapid Prototyping and Field Evaluation of Mobile Experiences. CHI '09 Workshop on *Mobile User Experience Research: Challenges, Methods & Tools*. April.

Bentley, Frank, and Crysta Metcalf. 2009b. The Use of Mobile Social Presence. *IEEE Pervasive Computing / IEEE Computer Society [and] IEEE Communications Society* 8 (4):35–41.

Bentley, Frank, Crysta Metcalf, and Gunnar Harboe. 2006. Personal vs. Commercial Content: The Similarities between Consumer Use of Photos and Music. In *Proceedings of the SIGCHI Conference on Human Factors in Computing Systems,* edited by R. Grinter, T. Rodden, P. Aoki, E. Cutrell, R. Jeffries, and G. Olson. Montréal, Québec. April 22–27. CHI '06. ACM, New York, NY, 667–676.

Bentley, Frank, Joe Tullio, Crysta Metcalf, Drew Harry, and Noel Massey. 2007. A Time to Glance: Studying the Use of Mobile Ambient Information. Pervasive 2007 Workshop on the Design and Evaluation of Ambient Information Systems. May.

Bernard, Russell H. 2002. *Research Methods in Anthropology: Qualitative and Quantitative Methods.* Walnut Creek, CA: AltaMira Press.

Beyer, Hugh, and Karen Holtzblatt. 1998. *Contextual Design: Defining Customer-Centered Systems.* San Francisco, CA: Morgan Kauffman.

Blackwell, Gerry. 2006. Mobile Messaging: Part II—The MMS Conundrum. Accessed February 20, 2011. http://itmanagement.earthweb.com/mowi/article.php/3607866.

Boost Mobile Where You At. 2011. Accessed February 27. http://www.juxtinteractive.com/work/boost-mobile-where-you-at/.

Borenstein, J. 2005. Camera Phone Images: How the London Bombings in 2005 Shaped the Form of News. *Georgetown University's Peer Reviewed Journal of Communication, Culture and Technology* 8 (2).

Brooks, Frederick P., Jr. 1995. *The Mythical Man-Month: Essays on Software Engineering.* Boston, MA: Addison Wesley.

Buechley, Leah. 2010. New Textiles Class Wearables Project. Accessed October 5. http://newtextiles.media.mit.edu/2010/pmwiki.php?n=Main.Wearable.

Bump. 2011. Accessed February 27. http://bu.mp/.

Büttner, Sebastian, Henriette Cramer, Mattias Rost, Nicolas Belloni, and Erik Erik Holmquist Lars. 2010. φ^2: Exploring Physical Check-Ins for Location-Based Services. In *Proceedings of the 12th ACM International Conference Adjunct Papers on Ubiquitous Computing.* Ubicomp '10. ACM, New York, NY, USA, 395–396.

Camtasia. 2011. Accessed February 15. http://www.techsmith.com/camtasia/.

Chen, Brian X. 2009. Verizon Leads, AT&T Runs Last in Wired.com's 3G Speed Test. Accessed January 8, 2011. http://www.wired.com/gadgetlab/2009/07/3g-speed-test/.

Chen, Mike Y., Timothy Sohn, Dmitri Chmelev, Dirk Haehnel, Jeffrey Hightower, Jeff Hughes, Anthony LaMarca, Fred Potter, Ian Smith, and Alex Varshavsky. 2006. Practical Metropolitan-Scale Positioning for GSM Phones. In *Proceedings of the Eighth International Conference on Ubiquitous Computing* (Ubicomp 2006). *Lecture Notes in Computer Science* 4206, 225–242.

Cheng, Y., Y. Chawathe, A. LaMarca, and J. Krumm. 2005. Accuracy Characterization for Metropolitan-Scale Wi-Fi Localization. In *Proceedings of the 3rd International Conference on Mobile Systems,*

Applications, and Services. Seattle, Washington. June 6–8. MobiSys '05. ACM, New York, NY, 233–245.

Church, Karen, and Mauro Cherubini. 2010. Evaluating Mobile User Experience In-The-Wild: Prototypes, Playgrounds and Contextual Experience Sampling. Presented at the Research in the Large: Using App Stores, Markets, and Other Wide Distribution Channels in UbiComp Research Workshop at Ubicomp 2010.

Clegg, Dai, and Richard Barker. 2004. *Case Method Fast-Track: A RAD Approach.* Boston, MA: Addison-Wesley.

Consolvo, S., D. W. McDonald, T. Toscos, M. Y. Chen, J. Froehlich, B. Harrison, P. Klasnja, et al. 2008. Activity Sensing in the Wild: A Field Trial of Ubifit Garden. In *Proceeding of the Twenty-Sixth Annual SIGCHI Conference on Human Factors in Computing Systems.* Florence, Italy. April 5–10. CHI '08. ACM, New York, NY, 1797–1806.

Consolvo, Sunny, and M. Walker. 2003. Using the Experience Sampling Method to Evaluate Ubicomp Applications. *IEEE Pervasive Computing / IEEE Computer Society [and] IEEE Communications Society* 2 (2):24–31.

Cramer, Henriette, Mattias Rost, and Lars Erik Holmquist. 2011a. Performing a Check-in: Emerging Practices, Norms and "Conflicts" in Location-Sharing Using Foursquare. In *Proceedings of the 13th International Conference on Human Computer Interaction with Mobile Devices and Services (MobileHCI '11).* Stockholm, Sweden.

Cramer, Henriette, Mattias Rost, and Lars Erik Holmquist. 2011b. Services as Materials: Using Mashups for Research. In *Proceedings of the 2nd International Workshop on Research in the Large* (LARGE '11). ACM, New York, NY, 9–12.

Deep Dive. 1999. *Nightline.* ABC News, July 13. Available from: http://abcnewsstore.go.com/ProductDisplay.aspx?ID=N990713.

Distimo.com. 2010. Apple iOS App Store, BlackBerry App World, Google Android Market, Nokia Ovi Store, Palm App Catalog, and Windows Marketplace. Accessed December 31. http://www.distimo.com/uploads/reports/Distimo%20Report%20-%20October%202010.pdf.

Distribute Your App. 2011. Accessed February 27. http://developer.apple.com/programs/ios/distribute.html.

Dougherty, Audubon. 2010. New Medium, New Practice: Civic Production in Live-Streaming Mobile Video. Master's thesis in Comparative Media Studies, MIT.

Dourish, Paul. 2006. Implications for design. In *Proceedings of the SIGCHI Conference on Human Factors in Computing Systems*, edited by R. Grinter, T. Rodden, P. Aoki, E. Cutrell, R. Jeffries, and G. Olson. Montréal, Québec. April 22–27. CHI '06. ACM, New York, NY, 541–550.

Ervin, Alexander M. 2000. *Applied Anthropology: Tools and Perspectives for Contemporary Practice.* Boston, MA: Allyn & Bacon.

Facebook Data Team. 2010. How Voters Turned Out on Facebook. Accessed January 8, 2011. http://www.facebook.com/note.php?note_id=451788333858.

FLIPOUT™ with MOTOBLUR™: A Stylishly Square & Pocket Perfect Mobile. Motorola launches in Brazil. Accessed February 27, 2011. http://mediacenter.motorola.com/Press-Releases/ Motorola-launches-in-Argentina-FLIPOUT-with-MOTOBLUR-Stylishly-Square-Pocket -Perfect-3254.aspx.

Fogg, B. J., and E. Allen. 2009. 10 Uses of Texting to Improve Health. In *Persuasive '09 Proceedings of the 4th International Conference on Persuasive Technology* 350: 1–6. Claremont, California. April 26–29. ACM, New York, NY.

Friedman, Max. 2011. Up Close with iOS5: Reminders. *MacWorld.com.* Accessed October 13. http://www.macworld.com/article/162991/2011/10/up_close_with_ios_5_reminders.html.

Garau, M., J. Poisson, S. Lederer, and C. Beckmann. 2006. Speaking in Pictures: Visual Conversation Using Radar. In: Second Workshop on "Pervasive Image Capture and Sharing: New Social Practices and Implications for Technology." (PICS) Ubicomp.

Gaver, Bill, T. Dunne, and E. Pacenti. 1999. Design: Cultural Probes. *Interaction* 6 (1):21–29.

Good, Nathaniel S., Jens Grossklags, Deirdre K. Mulligan, and Joseph A. Konstan. 2007. Noticing Notice: A Large-Scale Experiment on the Timing of Software License Agreements. In *Proceedings of the SIGCHI Conference on Human Factors in Computing Systems.* CHI '07. ACM, New York, NY, USA, 607–616.

Goodman, L. A. 1961. Snowball Sampling. *Annals of Mathematical Statistics* 32:148–170.

Google Apps Marketplace Terms of Service. 2011. Accessed February 27. http://www.google.com/ enterprise/marketplace/tos.

Google Gears Geolocation, A. P. I. 2010. Accessed October 5. http://code.google.com/apis/gears/ api_geolocation.html.

Google Gmail Mobile. 2011. Accessed February 27. http://m.gmail.com.

GPS & Selective Availability Q&A. 2010. Retrieved May 28. http://ngs.woc.noaa.gov/FGCS/info/ sans_SA/docs/GPS_SA_Event_QAs.pdf.

Grimes, Andrea, and Richard Harper. 2008. Celebratory Technology: New Directions for Food Research in HCI. In *Proceedings of the Twenty-Sixth Annual SIGCHI Conference on Human Factors in Computing Systems* (CHI '08). ACM, New York, NY, 467–476.

Grinter, Rebecca E., Leysia Palen, and Margery Eldridge. 2006. Chatting with Teenagers: Considering the Place of Chat Technologies in Teen Life. *ACM Transactions on Computer-Human Interaction* 13 (4):423–447.

Grossman, L. 2009. Iran's Protests: Why Twitter Is the Medium of the Movement. *Time Magazine,* June 7.

Harboe, Gunnar, Frank Bentley, Crysta Metcalf, and Vivek Thakkar. 2008. Method and System for Generating a Play Tree for Selecting and Playing Media Content. U.S. Patent 7,685,154.

Harboe, Gunnar, N. Massey, Crysta Metcalf, D. Wheatley, and G. Romano. 2006. Perceptions of Value: The Uses of Social Television. In *Proceedings of EuroITV 2007*, 116–125.

Henderson, Cal. 2006. *Building Scalable Web Sites: Building, Scaling, and Optimizing the Next Generation of Web Applications*. Sebastopol, CA: O'Reilly Media.

Hightower, Jeffrey, Sunny Consolvo, Anthony LaMarca, Ian Smith, and Jeff Hughes. 2005. "Learning and Recognizing the Places We Go." In *Proceedings of the Seventh International Conference on Ubiquitous Computing* (Ubicomp 2005), 159–176. September.

InfoTrends/CAP Ventures Releases Worldwide Mobile Imaging Study Results. 2011. Accessed February 27. http://www.infotrends.com/home/Press/2005/1.10.05.a.html.

Ion, Florence. 2010. Real Science Settles the Smartphone Display Wars. http://www.maclife.com/article/feature/real_science_settles_smartphone_display_wars.

iTunes Store Terms and Conditions. 2011. Accessed February 27. http://www.apple.com/legal/itunes/us/terms.html.

Jacob, Robert J. K., and Keith S. Karn. 2003. Eye Tracking in Human–Computer Interaction and Usability Research: Ready to Deliver the Promises. In *The Mind's Eye: Cognitive and Applied Aspects of Eye Movement Research*, ed. Jukka Hyönä, Ralph Radach, and Heiner Deubel. Oxford, England: Elsevier.

Juhlin, Oskar, Arvid Engström, and Erika Reponen. 2010. Mobile Broadcasting: The Whats and Hows of Live Video as a Social Medium. In *Proceedings of the 12th International Conference on Human Computer Interaction with Mobile Devices and Services*. MobileHCI '10. ACM, New York, NY, USA, 35–44.

Kasten, O., and M. Langheinrich. 2001. First Experiences with Bluetooth in the Smart-Its Distributed Sensor Network. Workshop on Ubiquitous Computing and Communications. PACT. October.

Kaufmann, Bonifaz. 2010. Amarino: "Android Meets Arduino". Accessed October 5. http://www.amarino-toolkit.net/.

Kay, Matthew, and Michael Terry. 2010. Communicating Software Agreement Content Using Narrative Pictograms. In *Proceedings of the 28th International Conference Extended Abstracts on Human Factors in Computing Systems*. CHI EA '10. ACM, New York, NY, USA, 2715–2724.

Kelley, Tom, Jonathan Littman, and Tom Peters. 2001. *The Art of Innovation: Lessons in Creativity from IDEO, America's Leading Design Firm*. New York, NY: Crown Business.

Kim, S., and E. Paulos. 2010. InAir: Sharing Indoor Air Quality Measurements and Visualizations. In *Proceedings of the 28th International Conference on Human Factors in Computing Systems*. Atlanta, Georgia. April 10–15. CHI '10. ACM, New York, NY, 1861–1870.

Last.fm Service. 2011. Accessed February 27. http://www.last.fm/.

Lewis, C. H. 1982. Using the "Thinking Aloud" Method in Cognitive Interface Design. Technical Report IBM RC-9265.

Lindlof, T. R., and B. C. Taylor. 2002. *Qualitative Communication Research Methods*. 2nd ed. Thousand Oaks, CA: Sage Publications.

Ling, Rich, and Brigitte Yttri. 1999. "Nobody Sits at Home and Waits for the Telephone to Ring": Micro and Hyper-coordination Through the Use of the Mobile Phone. Report No. 30/99, Telenor Research and Development, Oslo.

Liu, Sophia B., Leysia Palen, Jeannette Sutton, Amanda L. Hughes, and Sarah Vieweg. 2008. In Search of the Bigger Picture: The Emergent Role of On-Line Photo Sharing in Times of Disaster. In *Proceedings of the 5th International ISCRAM Conference.* Washington, DC. May.

Loopt. 2012. Accessed February 15. http://www.loopt.com/.

Lund, A. M. 1997. Expert Ratings of Usability Maxims. *Ergonomics in Design* 5 (3):15–20.

Manifest.permission | Android Developers. 2010. Accessed December 31. http://developer.android.com/reference/android/Manifest.permission.html.

McMillan, Donald. 2010. iPhone Software Distribution for Mass Participation. Presented at the Research in the Large Workshop, Ubicomp 2010. Copenhagen, Denmark.

Metcalf, Crysta. 2011. Using Ethnography for New Product Ideation at Motorola: Two Case Studies. Accessed February 27. http://www2.uta.edu/mpeterson/Qualitative%20Research/Metcalf_Crysta.pdf.

Method Cards for IDEO. 2002. http://www.ideo.com/work/method-cards. Accessed January 8, 2011.

Milgram, P., and A. F. Kishino. 1994. Taxonomy of Mixed Reality Visual Displays. *IEICE Transactions on Information and Systems. E (Norwalk, Conn.)* 77-D (12):1321–1329.

Miluzzo, Emiliano, Nicholas D. Lane, Shane B. Eisenman, and Andrew T. Campbell. 2007. CenceMe: Injecting Sensing Presence into Social Networking Applications. In *Proceedings of the 2nd European Conference on Smart Sensing and Context,* edited by Gerd Kortuem, Joe Finney, Rodger Lea, and Vasughi Sundramoorthy. EuroSSC'07. Berlin/Heidelberg: Springer-Verlag, 1–28.

MLB At Bat. 2010. Accessed December 31. http://mlb.mlb.com/mobile/atbat/.

Moore, Geoffrey A. 1991. *Crossing the Chasm: Marketing and Selling High-Tech Products to Mainstream Customers.* New York, NY: Harper Business Essentials.

Morrison, Alistair, and Matthew Chalmers. 2011. SGVis: Analysis of Data from Mass Participation Ubicomp Trials. *International Journal of Mobile Human Computer Interaction* 3 (4), 36–54.

MOTOACTV. 2012. Accessed January 7. http://www.motorola.com/motoactv.

Naaman, M., and R. Nair. 2008. ZoneTag's Collaborative Tag Suggestions: What Is This Person Doing in My Phone? *IEEE MultiMedia* 15 (3):34–40.

Naaman, M., R. Nair, and V. Kaplun. 2008. Photos on the Go: A Mobile Application Case Study. In *Proceeding of the Twenty-Sixth Annual SIGCHI Conference on Human Factors in Computing Systems.* Florence, Italy. April 5–10. CHI '08. ACM, New York, NY, 1739–1748.

Nakhimovsky, Yelena, Dean Eckles, and Jens Riegelsberger. 2009. Mobile User Experience Research: Challenges, Methods & Tools. In *Proceedings of the 27th International Conference Extended Abstracts on Human Factors in Computing Systems*. CHI '09. ACM, New York, NY, USA, 4795–4798.

Nakhimovsky, Yelena, Andrew T. Miller, Tom Dimopoulos, and Michael Siliski. 2010. Behind the Scenes of Google Maps Navigation: Enabling Actionable User Feedback at Scale. In *CHI 2010 Extended Abstracts on Human Factors in Computing Systems*.

Netflix. 2011. Accessed February 27. http://www.netflix.com.

Nielsen, Jakob. 1993. *Usability Engineering*. San Francisco, CA: Morgan Kauffman.

Nielsen, Jakob. 2000. Why You Only Need to Test with 5 Users. Alertbox, March.

Nielsen, Jakob. 2001. First Rule of Usability? Don't Listen to Users. Accessed August 3, 2011. http://www.useit.com/alertbox/20010805.html.

Nielsen, Jakob. 2003. Usability 101: Introduction to Usability. Accessed February 24, 2011. http://www.useit.com/alertbox/20030825.html.

Nielsen, Jakob. 2005. Ten Usability Heuristics. Accessed January 12, 2012. http://www.useit.com/papers/heuristic/heuristic_list.html.

Nielsen, Jakob, and H. Loranger. 2006. *Prioritizing Web Usability*. Berkeley, CA: New Riders.

nike+. 2012. Accessed January 7. http://nikeplus.com.

Noguchi, Y. 2005. Camera Phones Lend Immediacy to Images of Disaster. *Washington Post*, July 8.

Norman, Donald A. 1990. *The Design of Everyday Things*. Cambridge, MA: MIT Press.

Norman, Donald A. 1998. *The Invisible Computer*. Cambridge, MA: MIT Press.

Norman, Donald A., and Stephen W. Draper. 1986. *User Centered System Design: New Perspectives on Human-Computer Interaction*. Hillsdale, NJ: Lawrence Erlbaum Associates Inc.

O'Brien, Shannon, and Floyd Mueller. 2007. Jogging the Distance. In *Proceedings of the SIGCHI Conference on Human Factors in Computing Systems*. San Jose, California. April 28–May 3. CHI '07. ACM, New York, NY, 523–526.

Oldenberg, Ray. 1989. *The Great Good Place: Cafes, Coffee Shops, Community Centers, General Stores, Bars, Hangouts, and How They Get You Through the Day*. New York, NY: Paragon Books.

Pandora. 2011. Accessed February 27. http://www.pandora.com.

PDA Net. 2010. Accessed December 31. http://www.junefabrics.com/android/.

Pinch Media/Flurry. 2011. Accessed February 27. http://www.flurry.com.

Qik. 2011. http://www.qik.com.

Real HDMI Unlocks Motorola Droid X's Crippled HDMI Functionality. 2011. Accessed February 27. http://www.androidpolice.com/2010/08/30/new-application-for-motorola-droid-x-enables -added-hdmi-functionality/.

Reddy, Sasank, Katie Shilton, Jeff Burke, Deborah Estrin, Mark Hansen, and Mani Srivastava. 2008. Evaluating Participation and Performance in Participatory Sensing. International Workshop on Urban, Community, and Social Applications of Networked Sensing Systems. UrbanSense08. Raleigh, North Carolina. Nov 4.

Rettig, Marc. 1994. Prototyping for Tiny Fingers. *Communications of the ACM* 37 (4):21–27.

Riegelsberger, Jens, and Yelena Nakhimovsky. 2008. Seeing the Bigger Picture: A Multi-Method Field Trial of Google Maps for Mobile. In Chi '08 Extended Abstracts on *Human Factors in Computing Systems*. CHI '08. ACM, New York, NY, USA, 2221–2228.

Roving Networks. RN-41 Spec Sheet. 2010. Accessed October 2. http://www.rovingnetworks.com/ resources/download/18/RN_41.

Santora, T. 2006. Webster Dictionary Declares 'Crackberry' Its Word of the Year. *New Orleans City Business*, November 20.

Savio, Nadav, and Jared Braiterman. 2007. Design Sketch: The Context of Mobile Interaction. In *Proceedings of Mobile HCI 2007*.

Scupin, Raymond. 1997. The KJ Method: A Technique for Analyzing Data Derived from Japanese Ethnology. *Human Organization* 56 (2):233–237.

Seidio Innocell 2800 mAh Battery for Motorola Droid. 2010. Accessed December 31. http://www .amazon.com/Seidio-Innocell-Battery-Motorola-Replacement/dp/B003FSTBPE.

Shirky, C. 2008. *Cognitive Surplus: Creativity and Generosity in a Connected Age*. New York, NY: Penguin Press.

Shostack, G. Lynne. 1984. Designing Services that Deliver. *Harvard Business Review* 62 (1): 133–139.

Skyhook Wireless. 2010. Accessed October 5. http://www.skyhookwireless.com/.

Skype. 2011. Accessed February 27. http://www.skype.com.

Smith, Ian, Sunny Consolvo, Jeffrey Hightower, Giovanni Iachello, Anthony LaMarca, James Scott, Tim Sohn, and Gregory Abowd. 2005. Social Disclosure of Place: From Location Technology to Communications Practices. In *Proceedings of the Third International Conference on Pervasive Computing* (Pervasive 2005). May.

Sohn, Timothy, Kevin Li, Gunny Lee, Ian Smith, James Scott, and William G. Griswold. 2005. Place-Its: A Study of Location-Based Reminders on Mobile Phones. *Seventh International Conference on Ubiquitous Computing* (Ubicomp 2005). Tokyo, Japan.

Sohn, Timothy, Alex Varshavsky, Anthony LaMarca, Mike Y. Chen, Tanzeem Choudhury, Ian Smith, Sunny Consolvo, Jeffrey Hightower, William G. Griswold, and Eyal de Lara. 2006. Mobility Detection Using Everyday GSM Traces. In *Proceedings of the Eighth International Conference on Ubiquitous Computing* (Ubicomp 2006), 212–224. September.

Spotify. 2011. Accessed February 27. http://www.spotify.com.

Starner, T., S. Mann, B. Rhodes, J. Levine, J. Healey, D. Kirsch, R. Picard, and A. Pentland. 1997. Augmented Reality Through Wearable Computing. *Presence (Cambridge, Mass.)* 6 (4):386–398.

Taking Screenshots on an Android-based Phone. 2011. Accessed February 27. http://downloadsquad.switched.com/2008/10/22/taking-screenshots-on-an-android-based-phone/.

The Mobile Millennium Project. 2012. Accessed February 15. http://traffic.berkeley.edu.

Wang, Bo. 2009. Should an iPhone App Developer Charge or Run Ads? (Galaxy Impact Case Study). TechCrunch. http://techcrunch.com/2009/03/22/should-an-iphone-app-developer-charge-or-run-ads-galaxy-impact-case-study/. Accessed January 15, 2012.

Weilenmann, Alexandra. 2003. "I Can't Talk Now, I'm in a Fitting Room": Availability and Location in Mobile Phone Conversations. *Environment and Planning A* 35 (9): 1589–1605. Special issue on Mobile Technologies and Space, ed. E. Laurier.

Weilenmann, Alexandra, and Peter Leuchovius. 2004. "I'm Waiting Where We Met Last Time": Exploring Everyday Positioning Practices to Inform Design. *NordiCHI* 2004:33–42.

Wodtke, Christina. 2002. *Information Architecture: Blueprints for the Web*. Indianapolis, IN: New Riders.

Wortham, Jenna. 2009. Cellphone Locator System Needs No Satellite. *New York Times*, May 31.

Wroblewski, Luke. 2009. Mobile First. Accessed February 24, 2011. http://www.lukew.com/ff/entry.asp?933.

Yeh, Tom, John J. Lee, and Trevor Darrell. 2008. Photo-Based Question Answering. In *Proceedings of the 16th ACM International Conference on Multimedia*. Vancouver, British Columbia. October 26–31.

Young, Indi. 2008. *Mental Models: Aligning Design Strategy with Human Behavior*. New York, NY: Rosenfeld Media.

Index

Page numbers followed by *f* or *t* refer to figures and tables, respectively.